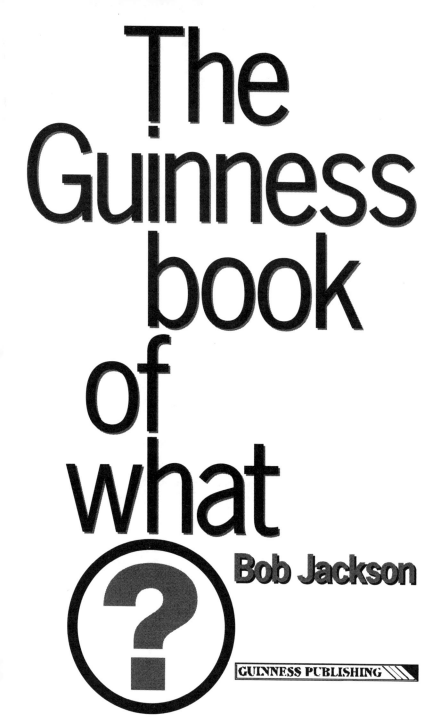

The Guinness book of what?

Bob Jackson

GUINNESS PUBLISHING

Editor: Beatrice Frei
Design and layout: John Mitchell, Mitchell Associates
Artwork: John Mitchell, except pages 50 and 216: Peter Harper

First published in 1995 by Guinness Publishing Limited
33 London Road, Enfield, Middlesex.

Printed and bound in Great Britain by Cox & Wyman Limited
A catalogue record for this book is available from the British Library.

ISBN 0-85112-684-7

Foreword

If you were to add up all the things in life you take for granted, you would probably lose count. For example, we all cry. We cry as small children in order to attract attention, because of some pressing need or hurt; we cry as adults in order to release emotion. But how often do we stop to think what natural mechanism enables us to cry; what is it that literally brings tears to our eyes?

This is just one of the many questions answered here. Others, involving matters of the human body, include topics such as depression, ulcers and other ailments ranging from malaria to the Black Death, the scourge of the Middle Ages. And if you remember the contents of this book after you have read it, will you pause to think about what chemistry lies behind the process we call memory?

Many questions concerning the nature of the planet Earth are also asked and answered, including some that are vital to its future. What, for instance, do scientists really mean when they speak of the greenhouse effect and global warming? Just what is the ozone layer, and what would happen to the world – and ourselves – if it no longer existed?

Most of us know the old rhyme *Red sky at night, shepherd's delight; red sky in the morning, shepherd's warning*. But what does it mean? What exactly is it that causes red skies at night and in the morning? You may think you know the answer to that one, but this book will tell you that red skies are not all they seem to be.

The Guinness Book of What? seeks to answer questions that are out of this world, too – some of which continue to mystify science. What, for example, is a black hole, and do such things really exist? What means do astronomers use to detect them, since they are completely invisible and their massive gravity prevents everything – including gravity – from escaping?

Come to think of it, exactly what is the force we call gravity? It allegedly caused Isaac Newton to think twice about sitting under an apple tree and it keeps our feet on the ground, but what creates it, and how does it fit into the universal scheme of things?

We can answer that, because we know the physical laws that govern gravity. Other questions, though, provide a good deal more food for thought. In science fiction tales, time travel is relatively routine – but will it ever become a reality? What futuristic machines will enable us to achieve it? You may be surprised to learn that scientists are seriously investigating the possibilities right now – and even putting forward theories on how a spacecraft might use the massive gravity of a black hole to hurl itself from one time to another.

Although the *Book of What?* concentrates on answering questions that arise from known scientific facts, it also investigates some of the more mysterious and esoteric aspects of the world about us. For instance, is a 'sixth sense' latent within us all? What exactly is it, and can we develop and manipulate it, as some animals seem to be able to do? Is it all tied in with manifestations of force such as gravity – and part of a deeper underlying force that lies behind the rest?

Perhaps the most intriguing questions of all posed in this book deal with Earth's mysterious past. What are ley lines – patterns crossing the landscape of Europe, said to follow lines of magnetic force between points of ancient significance? Are they part of an ancient knowledge long since lost – a knowledge that surfaces unknowingly in such old arts as dowsing, or water divining?

If, after reading this book, you are left with the feeling that what science dismisses today it will accept tomorrow, don't be surprised. That's what it is all about.

Contents

What is memory?

Of all the amazing functions of the human brain, that of memory is perhaps the hardest to understand. One of the reasons for this is that we undoubtedly take memory for granted, and consequently tend to overlook its power. Another reason is that memory does not seem to operate to any particular kind of order; sometimes our recollections of events and places are vivid and sharp, sometimes they are blurred and murky.

But memory remains one of the most potent forces in our lives from the moment of birth. A child learns new words at an average rate of ten a day, and by adulthood may have amassed a vocabulary of 100,000. The ancient Celtic and Scandinavian bards knew over 300 epic poems by heart, and performed a vital function in handing down the oral traditions of their peoples.

According to the two main existing theories, memory mechanisms operate through the medium of protein molecules or large collections of nerve cells called neurons. A neuron consists of a cell body, branchlike filaments called dendrites and a long shaft called the axon. To communicate with another cell, a neuron's cell body fires an electrical signal, the 'action potential', down the axon, triggering the release of neurotransmitter molecules. These travel across a gap, the synapse, to specialized receptors on the receiving neuron's dendrites.

Just how the memory stores the information that passes between the brain's neurons, however, remains something of a mystery. Scientists now think that recently acquired information is stored in the hippocampus, a structure deep within the brain that is thought to have much to do with human emotion, while the cortex – the outer layer of the brain – plays its part in storing older memories and is also essential for 'working' memory. Without this you would be incapable of holding a conversation because it enables you to hold fleeting material in your head so that you can build and understand complex sentences. This aspect of memory also comes into action if you need, say, to memorize a telephone number briefly before writing it down.

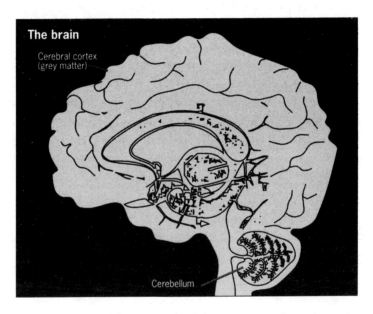

The brain

Cerebral cortex (grey matter)

Cerebellum

Neuropsychologists have identified three main mental components in working memory: the phonological loop is the machinery that enables us to retain a sequence of digits, letters or words; the visuo-spatial scratch pad is a kind of inner eye that lets us perform such functions as rotating shapes in our head, or remembering where we were on a printed page; and the central executive makes it possible for us to perform a great many tasks such as reasoning and mental arithmetic.

Memory, in effect, is like a building made up of many bricks, each one containing its own nucleus of information created by different learning processes. For example, semantic memory governs your knowledge of language and the world about you; episodic memory is your recollection of circumstances and events that have had a bearing on your everyday life; and classical conditioning is the way our memory tells us to respond to matters of routine, in much the way that Pavlov's celebrated dogs were conditioned to respond to a bell rung just before their feeding time. Experiments have shown that this type of memory involves the cerebellum, which is located at the back of the brain.

Modern research indicates that the way information is stored depends upon the way it was learned in the first place. Short-term memory (STM), our simplest memory-storage receptacle, serves as a kind of holding pen for data we may or may not wish to retain, while long-term memory (LTM) has a comparatively limitless capacity and duration. In order for information to make the leap from STM to LTM, it must have some significance or association. For instance, a car and its number plate glimpsed at random might soon be forgotten – but if you were to see the same car racing away from the scene of a robbery, you would probably remember it for the rest of your life.

What is schizophrenia?

Do the roots of schizophrenia lie in our genes, our environment or some combination of the two?

People who suffer from the most severe forms of mental illness, of which this is one, are termed psychotic. They lose all touch with reality and are affected by delusions, often believing themselves to be someone of great importance. On the other hand, the delusion may take the form of a persecution mania in which a person believes that he or she is being hounded by another person or even an organization, such as MI5.

A psychotic person also suffers from hallucinations; these can affect any of the senses, but hearing voices is the most common manifestation.

Cases of extreme psychosis have been well documented throughout history. One of the most famous is the biblical story of Nebuchadnezzar, who crawled around on his hands and knees eating grass. Shakespeare frequently used madness in the plots of his plays; the question of whether Hamlet was insane, or merely simulating madness, has plagued generations of schoolchildren, while Robert Louis Stevenson used the mental disorder multiple personality as the basis for his story of *Dr Jekyll and Mr Hyde*.

Until the end of the 19th century, all mental illnesses were classed simply as madness. It was a psychiatrist named Emil Kraepelin who divided them into 'manic depressive insanity' and what he called 'dementia praecox' – later to be known as schizophrenia.

About 30 per cent of patients suffering from schizophrenia make a full recovery after treatment and experience no further trouble. Others need regular treatment to hold the illness in check, while less than ten per cent fail to respond at all to any form of treatment and are condemned to suffer delusions and hallucinations for as long as they live.

One of the puzzling things about schizophrenia is that, unlike other forms of mental illness, it does not usually grow worse with age. Why it develops in the first place has long been a mystery, too, although researchers now think that the onset of the illness may be triggered by some fault in the brain that makes the sufferer unduly sensitive to stress. Whereas most people have learned to live with stress and to cope with it, some are thrown completely off balance by a stressful event and develop schizophrenia.

Investigations into the possible link between stress and schizophrenia seem to indicate that the illness is induced when stress occurs as the result of a sudden shock, such as a family bereavement or the unexpected loss of a job. Studies carried out by health authorities in several countries have shown that in many cases, people developing schizophrenia – or relapsing into the illness after apparently having been cured – suffered highly stressful events two to three weeks before its onset.

One important feature of schizophrenia is that it runs in families, and at one time there was a strong theory that a schizophrenic child was the product of schizophrenic or neurotic parents. This led to a widespread belief that schizophrenia developed as a result of the way in which a child was treated in its home environment, rather than as the result of a genetic abnormality.

However, current research is turning that theory upside down. Some scientists, using the latest brain imaging techniques, believe that the trouble starts in the hippocampus – the brain's emotional centre, already thought to have a bearing on the memory process – and that a child may 'catch' schizophrenia while still in the womb. They have found that in severe schizophrenia, many nerve cells in the hippocampus face the wrong way, sometimes by as much as 180 degrees, and that the amount of 'twist' tends to match the severity of the illness. This theory holds that schizophrenia arises during the early months of pregnancy due to a genetic defect or, perhaps, a viral infection. The embryonic cells rotate while glueing themselves to the forming hippocampus; the glue dissolves and the cells shift.

The out-of-line cells then act as a time bomb, exploding into the full-blown illness under conditions of severe emotional stress, perhaps decades later.

What is the aurora?

For thousands of years, people who live in the regions close to the Earth's poles have been awed and mystified by the aurorae – the glowing, shifting bands of light that dance across the winter sky.

Today, scientists know a great deal about the aurorae. But the lights – which are named after Aurora, the Roman goddess of the dawn – still retain some of their mystique.

The Aurora Borealis, or Northern Lights, present a far more spectacular display than their southern counterparts, the Aurora Australis. Although they are often visible in British skies, you need to travel to latitudes higher than 60 degrees North to see the Northern Lights at their best.

The start of an auroral display may often be mistaken for the rays of the rising Sun. But then the faint haze of light swells into yellow-green ribbons tipped in red that shiver across the sky like a great luminous curtain. The awesome display is caused by streams of electrons and protons, the so-called solar wind that sweeps across space from the Sun to collide with the Earth's upper atmosphere, exciting its molecules to higher energy levels. This excess energy is released in cascading sheets of colourful light.

Now, after years of research using sounding rockets and satellites, the aurorae have begun to give up their secrets. The cost of this research has been justified, because the aurorae do much more than provide a colourful picture show. They also play havoc with the ionosphere – the layer of charged particles in the upper atmosphere – and seriously distort short-wave radio transmissions. In addition, they can produce false echoes on radar equipment monitoring the flow of airline traffic over the North Pole.

Aurorae invariably result when jets of matter called solar flares erupt from the Sun at a time of intense sunspot activity. When the high-energy particles reach the Earth's atmosphere, the resulting aurora can set up a huge surge of electrical current. In 1972, for example, an aurora generated during a powerful series of solar flares caused a 230,000-volt transformer to explode in British Columbia.

Strange things can also happen in the upper atmosphere when the aurora heats up the rarefied gases that compose it. Wind velocities can increase to as much as 900 mph (1450 km/h), and the temperatures of atmospheric particles can reach 2000°F (1090°C).

What scientists are now seeking to establish is how all this auroral activity at altitudes of between 40 and 600 miles (60–960 km) can influence conditions on the Earth's surface. One thing they have learned is that the electrical impact of the aurora increases the density of nitric oxide particles at high altitude; these then descend to lower levels and diminish the ozone layer, which protects us from harmful solar radiation.

Growing numbers of scientists now think that the aurora acts rather like a trigger, setting off a particular weather pattern. In recent years, they have learned a lot by shooting sounding rockets into the heart of auroral displays at heights of 150 miles (250 km). The rockets' payloads carry magnetometers, optical scanners and particle counters.

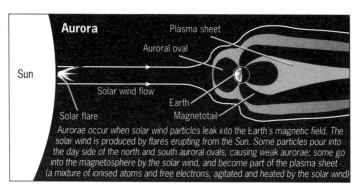

Aurora

Plasma sheet

Auroral oval

Sun

Solar wind flow

Earth

Solar flare

Magnetotail

Aurorae occur when solar wind particles leak into the Earth's magnetic field. The solar wind is produced by flares erupting from the Sun. Some particles pour into the day side of the north and south auroral ovals, causing weak aurorae; some go into the magnetosphere by the solar wind, and become part of the plasma sheet (a mixture of ionised atoms and free electrons, agitated and heated by the solar wind)

Artificial aurorae can also be formed by exploding metallic barium into the upper atmosphere, rather like dropping dye into a stream. The barium is ionized by solar radiation and begins to glow bluish-purple. Scientists can then observe how the glowing cloud spreads out on the magnetic currents at the edge of space - currents that can provide vital clues about the development of the weather systems that affect us all.

What are dreams?

Every night, each one of us enters into a mysterious, often terrifyingly real world – the world of dreams. For centuries, people have believed that this dreamworld holds the key to their personality and to their destiny. Yet the study of dreams, as a science, is comparatively new.

Not one of us can truthfully claim 'I never dream'. We all dream, every night; often several times in the course of a single night. In fact, dreaming accounts for 20 to 25 per cent of our normal sleep period.

It was two American neurophysiologists, Dr William Dement and Dr Nathanial Kleitman, who discovered that a person's eyeballs move very rapidly in the course of dreaming, and that this rapid eye movement – REM – could be registered electronically. They also discovered that a person who is awakened immediately after dreaming can recall dreamworld experiences in great detail, and that the time it takes to recount those experiences is exactly the duration of the dream itself – disproving an earlier theory that a dream never lasts for more than a few seconds.

As to why we dream, there are a number of theories. One of the most popular is that the brain operates selectively, scanning the mass of detail that bombards it during waking hours and immediately discarding unwanted material. But the brain also needs a period of consolidation when it can give its full attention to clearing the backlog, and this is what takes place during sleep. The process has been compared to a mainframe computer that goes 'off line' during the night and searches its files and programs, updating and modifying them to incorporate new data and erasing unwanted or redundant material.

Not all sleep theoreticians agree with this theory; many believe that the brain registers and stores everything, and is capable of recalling the most minute detail years after the event.

All researchers, however, are agreed on one matter; dreaming is good for us, and may even be a vital function. One of the first scientists to advance this theory was Dr Michel Jouvet, of Lyon, France. In one experiment, he attached electrodes to the base of a sleeping cat's skull, linking the animal with an apparatus known as an oneirograph. Whenever the cat began to dream, the oneirograph traced a vertical line on a revolving band of paper. As soon as this happened, Jouvet administered a slight electric shock to the animal – not enough to wake it, but enough to interrupt its dreaming. He observed that the cat began to spend more and more time in sleep, and that its dreams increased in frequency until they were almost incessant. Normally, a sleeping cat will dream at intervals of about 30 minutes, but after 24 hours of Jouvet's experiment the animal was dreaming continuously. Jouvet's conclusion was that dreaming performed a vital function in everyday life. Further evidence of this was provided by one of Dr Dement's experiments, in which several sleeping persons were awakened just as they started to dream. During the whole of the following day they were depressed, irritable and unable to concentrate. When the experiment was repeated without interruption for a fortnight, the subjects came dangerously close to a nervous breakdown.

Thanks to research of this nature, scientists have been able to draw up a typical programme for the average person's night's sleep. We begin by sleeping deeply for about one hour, and after a brief phase of relatively light slumber, deep sleep takes over again for another hour and a half. This is followed by a period of vigorous eyeball movement lasting approximately 20 minutes. Gradually, as the night goes on, the sleeper's mental activity is taken over more and more by dreams. Many dreams last for about an hour, and in some cases longer. On average, we usually dream four times during an eight-hour sleep period, each dream being somewhat longer than the one before.

What makes us left-handed?

Society is more enlightened about such things nowadays, of course, but there is no denying that left-handers had a hard time in days gone by. At best, they were looked upon as abnormal, while in mediaeval times they were thought to be tainted by the devil. Within living memory, left-handers were often forced to write with their right hand at school, with the predictable consequence that people found great difficulty in distinguishing what they wrote thereafter.

Devil-marked or not, there have been some celebrated left-handers in history. They include Alexander the Great, Horatio Nelson, former US President George Bush and Prince Charles, not to mention this writer. But what actually causes left-handedness is still a matter of some controversy.

Left-handedness runs in families. Experts have calculated that the chance of having a left-handed child is two to six per cent for two right-handed parents, up to 15 per cent where one parent is left-handed, and 30 to 50 per cent for two left-handed parents.

Researchers are keen to learn more about left-handedness, since it may hold the key to certain aspects of how the brain functions. Each half of the brain controls the opposite side of the body, plus certain specialized functions such as speech (the left side) and spatial ability (the right side). Clearly, left-handedness is more likely if the right hand cannot be used due to some physical infirmity or disorder, or because of some neurological condition, such as epilepsy, dyslexia, or autism.

Some researchers think that left-handedness may be the result of some imbalance that affects the brain before birth – perhaps an unusually high level of testosterone, the male hormone. This might explain why some surveys have established that there are more left-handed men than women.

There are, however, some puzzling aspects about the condition. For example, it is three times as common among premature babies, and more common among high-IQ groups, including precocious children with a high mathematical, musical or verbal ability.

One theory for this is that the genetic bias which, in the case of most people localizes language skills on the left side of the brain, also causes them to be right-handed. If the gene is absent, however, left-handedness might occur at random.

According to some researchers, right-handers tend to use the left side of their brain at the expense of the right. The brains of left-handed people have much less specialization on either side, and with ambidextrous folk the spread is pretty well even.

And there's a myth about left-handers that needs to be destroyed – they are not, and never have been more accident-prone than their right-handed brethren.

What is our body clock?

The experience is familiar to any air traveller who has crossed several time zones in a single day. While the sun may tell us it's 6.00 pm, our bodies disagree, insisting that it's midnight and time for bed. Such confusion, known as jet lag, is caused by the temporary disorientation of what is popularly called the body clock.

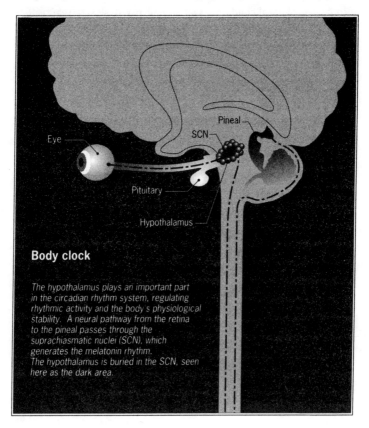

Body clock

The hypothalamus plays an important part in the circadian rhythm system, regulating rhythmic activity and the body's physiological stability. A neural pathway from the retina to the pineal passes through the suprachiasmatic nuclei (SCN), which generates the melatonin rhythm. The hypothalamus is buried in the SCN, seen here as the dark area.

Left to its own devices, the body clock completes its cycle over a period of about – but not exactly – 24 hours, hence its correct title of circadian rhythm (from the Latin *circa*: about, and *diem*: day). In this free-running state, all that drives the clock is the activity of its constituent nerve cells. In normal, everyday life we never experience this state, because the clock is conditioned by time cues in the environment; these synchronize the clock with the daily solar cycle.

Daylight, however, is only one of many potential time cues, known to chronobiologists as *Zeitgeber* (the German word meaning 'time givers'). For example, our regular cycle of sleep and activity affects the times when we eat, our social and work activities and our exposure to daylight.

It has been known for a long time that seasonal changes in the length of the day influence the biological rhythms of animals, and since the mid-1980s we have also known about something called Seasonal Affective Disorder (SAD) in humans, a mild form of depression normally associated with the winter months. Scientists are now fairly certain that this is brought on by the circadian rhythm's response to daylight, or rather the lack of it.

There are two main components of the circadian rhythm system, at least one of which is controlled in the brain by the hypothalamus. This is the region of the brain below the cerebrum which regulates rhythmic activity and physiological stability within the body, including water balance and temperature; it also regulates the production of the pituitary gland's hormones and controls that part of the nervous system regulating the involuntary muscles. In normal persons, the body clock's components are both synchronized with the sun's cycles, but in the case of people with depression or manic depression, one of them is out of synch.

One part of the circadian clock controls the timing of the sleep-wake cycle, together with levels of growth hormones and other substances. The second component regulates temperature and REM (rapid eye movement) sleep, the period when we dream. It is this part of the internal clock that goes awry in jet lag and manic-depressive disorders. Psychiatrists have been able to draw certain parallels between a depressed person's circadian rhythms and those of an airline passenger travelling over several time zones. The traveller's sleep schedule shifts to the new time zone and he goes to bed later than normal, but his REM-temperature system does not adjust for several days.

In the 1980s, psychiatrists at the US National Institute of Mental Health did some tests with depressed patients in which their sleep was moved forward by six hours. Their depression immediately lifted, and they retained normal moods for up to three weeks before the abnormal timing re-established itself. The researchers advanced the theory that the depressed patients were unusually sensitive to light; this sensitivity was transmitted by a portion of the optic nerve to the hypothalamus, which in turn affected the circadian rhythm.

Just what it is that makes the circadian clock tick in the first place is still something of a puzzle, but one theory suggests that it may be activated by a hormone which, in the first instance, is triggered by the reception of light via the optic nerve in the suprachiasmatic nuclei – the innermost region of the hypothalamus. Each no bigger than a grain of sand, these nuclei, called SCN for short, lie at the base of the brain near the optic tract. They are known to act as receptors for the hormone melatonin, which has long been associated with circadian rhythms.

What is a diamond?

Diamonds, as is well known, are a girl's best friend. They are also hard enough to tip drills and cutting tools. In fact, diamond is the hardest substance known. On the Mohs' scale (named after the German-Austrian mineralogist Friedrich Mohs), which provides a scale of hardness for minerals in ascending order, diamond ranks as No 10 after talc, gypsum, calcite, fluorite, apatite, orthoclase, quartz, topaz and corundum. Even then, the scale is not regular; diamond is 90 times harder in absolute terms than corundum, the mineral at No 9.

Diamonds were known as gemstones more than 5000 years ago. They are described in the Sanskrit records of ancient India, and until they were discovered in Brazil in 1725 India remained the principal source of supply. The main present-day sources are Australia, Zaïre, Botswana, the Yakut region of the Russian Federation, South Africa, Namibia and Angola.

Apart from its hardness, diamond has very attractive visual characteristics caused by its high refractive index (2.42) and high dispersion of light, the 'fire in the stone'. Largely because of this, magical powers were once attributed to it; in mediaeval times, crushed diamond was administered as a medicine to Europe's nobility, which doubtless did their stomachs no good at all.

It was not until the end of the 18th century that anyone made a serious study of the nature of diamond. Then, in 1798, a British chemist called Smithson Tennant succeeded in burning diamond and weighing the carbon dioxide that this process produced. His conclusion, subsequently proved by others, was that diamond was a form of carbon, like graphite.

In fact, diamond consists of a dense array of carbon atoms joined by strong bonds arranged in the form of a tetrahedron – a four-faced figure contained by four triangles – arranged around each atom. Graphite, on the other hand, has a structure similar to chicken wire, consisting of hexagonal layers of carbon atoms joined together by stronger bonds within the layers and weaker forces between them. This arrangement allows the layers to slide over each other, which gives graphite a rather greasy feel.

Diamond is formed under conditions of extreme temperature and pressure from carbon-bearing rocks very deep in the earth, near the boundary between crust and mantle. The resulting material is brought to the surface mainly by kimberite, an igneous rock found in carrot-shaped, pipelike intrusions called diatremes, where mobile material from very deep in the Earth's crust has forced itself upwards, expanding as it ascends. Most of the world's diamonds are mined from kimberite, but may also be found as alluvial deposits on or close to the Earth's surface in dried riverbeds or watercourses.

Rough diamonds are often dull and greasy before being sawn and then polished, using a mixture of oil and diamond powder. Gem diamonds are valued by weight, cut, colour and clarity; industrial diamonds are produced synthetically from graphite, a process first investigated in the former Soviet Union before the Second World War.

What is terraforming?

The human race can take heart: when the poor old Earth is worn out, poisoned and suffocated, we'll go off and build another one somewhere else.

Ever since space probes revealed that Mars was an icy, crater-strewn desert swept by fearsome winds and that Venus was a 1580°F (860°C) furnace with an atmosphere saturated with sulphuric acid, a kind of cold despair has gripped scientists who clung to the belief that there might be some form of organic life in other parts of our solar system. Apart from Earth, the worlds of the solar system are dead or totally inhospitable – but it might not always be so, if a new and incredibly futuristic science ever comes to reality.

Known as terraforming, the new science envisages a time in the future when scientists will be able to apply technology to the planets and turn them into habitable places.

It all sounds very far-fetched at the moment, but leading scientists at America's space agency, NASA, do not think so. As long ago as 1975, NASA stated that the field of terraforming was well worth serious study, and four years later a major conference on the possibilities of planetary engineering was held at Houston, Texas.

Already, preliminary studies have taken place on the feasibility of providing the Moon with an atmosphere. One possible method would be to vaporize parts of the lunar surface with controlled thermonuclear explosions; this would surround the Moon with a gaseous envelope which could then be further engineered to contain oxygen, nitrogen and hydrogen.

NASA scientists have also studied the possibility of 'kicking' an asteroid out of its orbit around the Sun and bringing it into orbit around the Earth so that its resources could be mined. Some asteroids are believed to be rich in precious metals and other elements that are now scarce on our own planet. Small asteroids could be knocked out of their orbits by controlled nuclear explosions. Another possible method would be to use a 'mass driver', a mechanism that would hurl huge chunks of the asteroid into space and produce a forward thrust in the opposite direction. The snag here would be that much of the asteroid – as much as three-quarters – would be consumed before the remainder reached its destination.

Some simple planetary engineering might be carried out with technology that could be developed in the foreseeable future. The amount of energy received by areas of a planet could, for example, be varied by concentrating sunlight onto them by means of huge mirrors in orbit. Such reflective surfaces would be up to 5 miles (8 km) in diameter and would be constructed of the lightest possible materials. These giant mirrors could also be moved across the solar system, from one planet to another, using energy provided by the 'solar wind' – cosmic particles streaming from the Sun, with enough energy to push the giant mirror assemblies through space.

The possibilities of terraforming are endless. One day, men will be able to construct artificial suns, blazing thermonuclear furnaces capable of sustaining themselves – in the same way as a natural star like the Sun – by the atomic forces unleashed within. When that happens, even the coldest and most distant planets could be provided with enough warmth and light to foster life.

Such artificial suns would enable men to colonize the larger satellites of the great 'gas giants', Saturn and Jupiter. Some of these satellites have their own internal heat sources; Io, for example, one of Jupiter's moons, has many active volcanoes, constantly erupting and propelling sulphurous materials to great heights above the satellite's surface at speeds of over 2000 mph (3000 km/h). One day, man will have the technological skill to harness this incredible energy.

The more scientists consider the potential of terraforming, the more they are reaching the conclusion that it may be the only ultimate way out for mankind. Sooner or later, as Earth's resources dry up, the human race must expand outwards into the universe – but even the nearer stars, with whatever planets orbit them, may be beyond our reach.

Over the next few centuries, terraforming could bring a workable solution – providing valuable resources and new homes in space.

What is Nemesis?

In 1984, a group of American scientists put forward a novel theory to explain why the Earth has apparently undergone cataclysmic upheavals at various times in its history.

They claim that every 26 million years or so, our planet is subjected to an intense bombardment by comets and other space debris – and it happens because our Sun has a dark, invisible companion star, orbiting far beyond the outer limits of the solar system.

The theory suggests that the dark star, which they named Nemesis after the Greek goddess of vengeance and divine retribution, goes round the Sun in an elliptical orbit at an average distance of about one light year. This is the distance covered by a ray of light in one year, travelling at 186,000 miles/sec (299,000 km/s). The Earth's distance from the Sun, by comparison, is a mere 92 million miles (148 million km).

The scientists, all researchers at the University of California, based part of their theory on the Earth's fossil record. At regular intervals over the past 150 million years – which is as far back as the fossil record goes with any accuracy – there seems to have been a massive extinction of various life forms, of which the dinosaurs are the most famous example.

This wholesale extinction of life on Earth – which appears to happen at intervals of between 26 and 30 million years – has long been the subject of heated scientific debate. According to one theory, it occurs because the Earth periodically undergoes a sudden and dramatic shift on its axis, like a gyro toppling off balance. This causes gigantic changes in the planet's climate, as well as massive volcanic upheavals, during which entire species may be wiped out.

The Nemesis theory, however, suggests a different cause. The scientists who propose it claim that the extermination of species coincides with a fierce bombardment from outer space. There is evidence that the Earth was hit by at least one, perhaps several, massive objects 65 million years ago, about the same time that the dinosaurs began to disappear, and there is doubtless fresh evidence waiting to be uncovered that similar cataclysms have happened since, although probably not on such a massive scale. The surface of the Earth is pock-marked with the huge scars of ancient craters, some only recently identified. The impact would have hurled millions of tonnes of dust and debris into the atmosphere. Combined with the smoke of newly erupted volcanoes, it would have formed a shroud around the Earth, blotting out the sunlight for months or even years. Plants would have died, and so would the creatures that fed on them. Only the more adaptable creatures – including the burrowing mammalian ancestors of mankind – would have been able to survive. Why did it happen? Because Nemesis, the dark star, makes its closest approach to the solar system every 26 million years or thereabouts, and in so doing disturbs a mass of comets orbiting far out in deep space.

According to the Dutch astronomer J.H. Oort, there is a 'cloud' of comets orbiting the Sun at a distance of about a light year. When

Nemesis passes this cloud, its gravity disturbs the comets and sends a shower of them plunging into the heart of the solar system. Some strike the inner planets – including Earth – with devastating effect.

Many astronomers believe the theory is absurd and have held it up to ridicule. Others, however, take it seriously and are examining the data gathered by astronomical satellites in an attempt to find some evidence that Nemesis exists. If it does, and if the 26-million-year orbital period theory holds good, then mankind could be in for a severe shock in about 13 million years' time.

What are the greenhouse effect and global warming?

These two aspects of the present-day terrestrial climate have probably received more media coverage than any other scientific matter in the late 20th century. They are treated together here because they are indivisible; one is a consequence of the other. The precise temperature at the surface of the Earth depends on a fine balance struck between incoming and outgoing energy. If carbon dioxide is added to the atmosphere, the balance is tilted towards warmth.

Carbon dioxide is a relatively minor constituent of the atmosphere, but it plays an important part in determining climate because of the way in which it absorbs infrared radiation. Most of the Sun's energy is radiated in the visible part of the spectrum and passes through the atmosphere to warm the surface of the Earth, but the warm ground and sea radiate part of this solar energy back into space at longer wavelengths, in the infrared. Some of this infrared radiation is absorbed by the carbon dioxide in the atmosphere and is re-radiated back down to the ground, so that the carbon dioxide is really acting in much the same way as the panes of a greenhouse, in that it hinders the escape of re-radiated heat back into space. In simple terms, the more carbon dioxide in the atmosphere, the more the surface temperature builds up to form the so-called 'greenhouse effect'.

The theory is by no means new; it was first postulated in 1827 by the French mathematician Joseph Fourier (1768-1830), and the term greenhouse effect was coined by the Swedish scientist Svante Arrhenius (1859-1927).

The concentration of carbon dioxide in the atmosphere has been building up steadily since the middle of the 19th century, when there was a sudden upsurge in human activity that involved the destruction of forests and the burning of fossil fuels, mainly coal. Since that time, according to the most accurate estimates, the concentration of carbon dioxide in the atmosphere has increased by around 25 per cent. In 1900, the first year when reliable measurements became available, the concentration was roughly 300 parts per million (ppm). By 1958, when very accurate measurements became available from the observatory on top of Mauna Loa, in Hawaii, the concentration had risen to 316 ppm, and by the mid-1980s it was 345 ppm.

The upward trend in carbon monoxide levels and the associated rise in temperature has been ongoing since 1850. Scientists predict that if the trend continues, temperatures will have risen by 37°F (3°C) by the end of the 21st century, making the world warmer than it has been for 100,000 years; and if human influences push carbon dioxide levels past the 350 ppm mark, the Earth will be in a regime of climate not experienced for a million years. There is also the frightening possibility that large amounts of carbon dioxide might then be released from the Earth's oceans as they warm up, producing a 'runaway' greenhouse effect.

Although the rise in carbon dioxide emissions through industry and other factors has been a primary cause of global warming since the mid-19th century, the biggest contributor to the greenhouse effect is in fact water vapour; clouds tend to become wetter and denser as warming causes increased evaporation, and denser clouds reflect more sunlight, at the same time trapping existing warmth close to the surface. Other gases such as methane, nitrous oxide, ozone and chlorofluorocarbons (CFCs), of which more later, have a part to play.

Not all scientists, however, are agreed on the global warming scenario. Some have even challenged the assumption of United Nations meteorologists that man-produced 'greenhouse gases' are the principal cause, claiming instead that variable solar radiation might be responsible. They say that the Sun, which has experienced a period of high activity during most of the 20th century, will calm down again in the 21st century, bringing cooler climatic conditions.

What is a tidal wave?

On 24 Apr 1771, a great wall of water more than 278 ft (85m) high appeared off Ishigaki island, part of the Ryukyu island chain in the Pacific. The great wave, the biggest ever recorded, tossed a 750-tonne block of coral more than 1.3 miles (2.5 km) inland. It was what the Japanese call a tsunami, a seismic tidal wave that results from an earth tremor or a volcanic disturbance deep down under the ocean bed.

For centuries, tsunami have been rolling across the world's oceans with devastating effect. Travelling at anything up to 500 mph (800 km/h), they have been known to cross thousands of miles of open sea, finally bursting on some stretch of coastline with the pent-up fury of an atom bomb.

In 1755, a tsunami wave – following closely on the heels of an earthquake – smashed into Lisbon. Together, the two catastrophes claimed 32,000 lives. Thirty years later, another giant wave killed 100,000 people along the coast of Calabria, in Italy. One of the worst tidal waves this century occurred in 1908, when the Sicilian town of Messina, already devastated by an earthquake, was hit, with the death toll rising to 86,000. And in 1923, a monstrous wave rolled into Sangami Bay near Tokyo, again in the aftermath of an earthquake, bringing the combined total of deaths to 100,000.

Today, although considerable progress has been made in the study of tidal waves and their origins, there are still wide gaps in our knowledge, and scientists from many countries are combining their efforts to try to create an efficient world-wide tidal wave early-warning system. Japan, where tsunami have caused more destruction over the years than anywhere else, was the first country to embark on an extensive tidal wave research programme, in 1933. In that year alone, the huge waves claimed 3000 Japanese lives and destroyed 4000 homes.

The Japanese established beyond all doubt that the tidal waves had two main causes; earth movements and volcanic action. They found that in the open sea, a tidal wave starts as a gentle swelling on the surface, about 3 ft (1 m) high by 80 miles (130 km) long. Only when it reaches shallow water does it build up into a towering mass of energy capable of enormous destruction. The average tsunami is so powerful that it may take more than a week to subside, rebounding from the coasts and re-crossing the ocean several times.

In the USA tidal wave research began with an unexpected eye-witness account of a disaster. On 1 Apr 1946, a strong earth tremor shook the sea bed off Alaska. Four and a half hours later, after crossing 2300 miles (3700 km) of ocean, the resulting tidal wave struck Hawaii, killing 173 people in the town of Hilo.

One man who survived the catastrophe was the American geologist Francis P. Shepard, who worked for the Scripps Institute of Oceanography in California. He found that on average, a really serious seismic tidal wave occurs once a year, mostly in the Pacific or the Indian Ocean, but occasionally – perhaps once or twice a century – in the Atlantic or Mediterranean.

To obtain new information, the scientists created their own miniature tidal waves off the Californian coast by detonating high explosives under water, and measuring the size and intensity of the resulting waves. As a consequence of their research, a seaquake early warning network was established around the Pacific to report any tremors or unusual disturbances under the sea. But although the warning system is as efficient as modern science can make it, man is still powerless to stop the tidal waves forming in the first place. That is a problem for the oceanographers of the future to tackle.

What is the ozone layer?

As if climatologists didn't have enough to worry about with the greenhouse effect and global warming, the 1980s identified another problem that could have a profound effect on life on Earth: the depletion of the ozone layer. This tenuous layer of gas, tens of miles above our heads, is an essential part of planet Earth's life support system; without it, it is doubtful whether there would be any life on land.

Ozone, a very active bluish gas, is formed when oxygen is subjected to an electrical discharge. At sea level it exists only in very small amounts, and the belief that you inhale large quantities of it at the seaside is an old myth; the health-giving qualities of sea air are due to a variety of other causes.

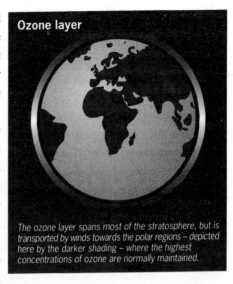

Ozone layer

The ozone layer spans most of the stratosphere, but is transported by winds towards the polar regions – depicted here by the darker shading – where the highest concentrations of ozone are normally maintained.

In the upper atmosphere, ozone is present at altitudes of between 9 and 25 miles (14 and 40 km), with the biggest concentration at the higher levels. In fact, the ozone layer comes more or less in the middle of three distinct regions of the upper atmosphere, and is responsible for an odd variation in temperature that takes place.

Up to an altitude of about 10 miles (16 km), in the region called the troposphere – from which weather forecasting information is gathered – the temperature falls steadily, sometimes as far as −112°F (−80°C). But higher up, in the stratosphere, it begins to rise again – a heating effect caused by the absorption of the Sun's ultra-violet radiation by the ozone. This increase in temperature occurs in a region known as the stratopause; above it, the temperature starts to plummet once more.

The ozone layer spans most of the stratosphere. It exists because oxygen filtering up from the top of the troposphere reacts under the influence of sunlight to form ozone. The concentration of ozone is greatest above the equator and the tropics because that is where solar radiation is strongest and most direct. From these regions, ozone is transported by winds throughout the stratosphere around the Earth towards the polar regions, so that the layer is maintained fairly uniformly – or so it was, before recent discoveries showed otherwise.

Without the ozone layer, the Earth's surface would be blasted by a degree of ultra-violet radiation that would be lethal to most forms of plant life, certainly the crops on which the human race depends for its survival. Even if humanity did manage to survive – and it probably would, given its built-in adaptability – it would have to live with widespread illnesses such as skin cancer and eye cataracts, all brought about by the increase in radiation from the Sun.

It was therefore a cause for serious concern when scientific observation in the mid-1980s revealed that a continent-sized hole had formed in the ozone layer over Antarctica, and in February 1988 another hole was located over the Arctic regions.

At first, it was thought that the ozone layer was being affected by the enormous quantities of dust and debris hurled into the upper atmosphere by a series of big volcanic eruptions that had recently taken place – natural catastrophes which had already combined to have a damaging effect on the Earth's climate. It is a fact that pollutants such as smoke and ash, if present in large enough quantities, can reduce the ozone level in affected areas by as much as ten per cent, and that the damage can last for weeks and even months.

Before long, however, scientists had tracked down the real culprit. During the last quarter of the 20th century, unnatural quantities of chlorine have been released into the stratosphere from chlorofluorcarbons – CFCs – which are chemicals used in aerosols, refrigerants and plastic foam. These drift up into the stratosphere and break down into chlorine atoms which become electrically charged by solar particles and act as catalysts to cause the ozone layer to break down.

This discovery sparked off an international effort to rescue the ozone layer by reducing the output of CFCs, but it is likely to be many years before the effects of such an operation become noticeable. But all the surrounding publicity has drawn worldwide attention to the fact that the ozone layer is a very good thing for life on Earth. After all, there are only about five billion tonnes of ozone in the entire stratosphere, and if it were all brought down to the surface, atmospheric pressure would squeeze it into a layer just three millimetres thick. That's a very modest amount of gas to do such an important job.

What are ley lines?

Science may soon be getting to grips with an age-old mystery that is right under our feet – the web of strange alignments, known as ley lines, that link holy places and sites of antiquity across the whole of Europe.

It was in 1921 that an amateur local historian named Alfred Watkins, standing on a high hilltop in Herefordshire, suddenly realized that all the ancient landmarks as far as the eye could see appeared to fall into a coherent pattern. When he marked out the churches and ancient sites on an Ordnance Survey map, he found that dozens of them lay in an arrow-straight alignment across the countryside as though they had been deliberately surveyed.

He also discovered that many of the places lined up in this way had common elements in their names. White, Black, Cole, Cold, Merry and Ley occurred most frequently, and from the last of these Watkins gave the mysterious alignments the name of ley lines.

Watkins' theory was that the lines across the landscape, some of which coincided with ancient tracks, had indeed been surveyed and constructed by some ancient race with a deep understanding of nature and astronomy, perhaps the same people as those who built Stonehenge.

The archaeologists of Watkins' time ridiculed his theory. Most of them believed that civilization had arrived in Britain with the Romans, and that the Celtic and pre-Celtic inhabitants of the British Isles had been little more than barbarians.

Today, we know otherwise. Archaeology is turning up evidence all the time to show that the Celts, and probably the civilizations that preceded them, were more advanced than the Romans in many respects. Not only were they expert craftsmen, farmers and foresters, they – or at any rate their priesthood – had a thorough knowledge of astronomy and a number of other sciences. And when the Romans laid down their network of military roads in Britain, it is now thought that their surveyors followed the lines of ancient roads that may have predated the Roman Empire by thousands of years.

Current researchers into the matter believe that ley lines are, in effect, navigational routes joining landmarks such as mounds, old stones, crosses and crossroads, moats, holy wells and churches. The lines, which run for many miles across the landscape in all directions, cross hilltops on which beacons probably once stood, and often end at a very high prominence such as a mountain peak.

The reason that churches are a feature of ley lines is that many of the early churches were built on the sites of pre-Christian temples. For centuries after Christianity came to Britain the dividing line between the new religion and the old ones was very thin, and building on the sites of pagan temples was one way of showing that the new religion was more powerful than anything that had gone before.

There has been a great upsurge of interest in ley lines in recent times, and fresh discoveries at Stonehenge and elsewhere have raised new questions. One recent theory suggests that Stonehenge and the lesser stone circles of western Europe are astronomically aligned to the whole ley line system, but the reason for this alignment is not yet clear.

One fact is now apparent; the stone circles are not haphazard. All are built to the same ground plan, using precise geometrical patterns which in turn are based on Pythagorean triangles – showing that whoever built them knew mathematical principles first thought to have been devised by the Greeks several thousand years later.

Modern research techniques may one day force the stone circles and ley lines to give up their secrets. But so far, everything points to the probability that our remote ancestors were masters of a scientific world whose treasures await rediscovery.

What is an iceberg?

There are two forms of ice in the waters of the Earth's polar regions: pack ice and icebergs. Pack ice forms from seawater and is recently created, rarely lasting more than a year or two; icebergs, on the other hand, are the fragments of ice sheets and glaciers that have built up on land over thousands of years.

Glaciers form when the amount of snow and freezing rain that is precipitated over a continent exceeds the degree of evaporation that takes place. Over the centuries, the glaciers gradually flow into the ice shelves that surround the polar continents, increasing the weight of the shelves and deforming them. This glacial movement, which can be several hundred or several thousand metres a year, is known as 'creep', the ice crystals continually sliding and stretching. This steady advance of the ice front, the seaward limit of the shelves, is interrupted from time to time; the build-up of weight becomes excessive and a lump breaks off. This is the birth, or 'calving' of an iceberg.

It seems a simple enough process, but in fact it is not. Various mechanisms go into the making of an iceberg, not the least of which is water pressure. At the point where glaciers or extensive ice shelves meet the sea, water pressure under the outer extremity of the glacier or shelf interacts with the main body of ice and the swell of the tides, which can be very marked in polar regions, exerting a force on the extremity that intermittently increases and decreases. The result, eventually, is that a large monolith of ice breaks away. But icebergs can form in other ways; sometimes, for example, the degree of melting or evaporation that takes place on the surface of a glacier near its seaward extremity is greater than the water erosion on its underside. This creates an underwater shelf which eventually breaks away and floats to the surface.

After calving, an iceberg can float for very long distances. Arctic icebergs can move thousands of miles in a few years, slipping down the eastern coast of North America to be caught up in the current of the Gulf Stream which can carry them, before they melt, to within a few hundred miles of the British Isles. However, of the 10,000 to 15,000 Arctic icebergs that calve annually, fewer than 400 pass Newfoundland to enter the North Atlantic.

Arctic icebergs are generally much smaller than their Antarctic counterparts, but large ones have been encountered over the years. In 1882, a monster measuring 7 miles long by 3.7 miles wide (11 × 5.9 km) was sighted near Baffin Land, and in the Second World War an Allied convoy nearly came to grief when it ran headlong into a berg measuring 4500 feet long by 3600 ft wide by 60 ft high (1370 × 1100 × 18 m) south of Newfoundland. Several ships collided as they tried to avoid it. Unlike the Titanic in the same waters in 1912, they succeeded.

The dimensions of icebergs become even more impressive when one considers that 80 per cent of their bulk lies under water. Antarctic icebergs are particularly massive; lengths of up to 5 miles (8 km) are not uncommon, with 150 ft (45 m) of ice visible above water. Some 93 per cent of the world's icebergs are produced in the southern polar regions.

In September 1986, three huge icebergs broke away from the Filchner Ice Shelf, about 400 miles (645 km) east of the Antarctic Peninsula. Together, they covered an area the size of Northern Ireland. Two went aground, but the third floated free in 1990 and headed north, giving scientists an unprecedented opportunity to track its movements. Over the years that followed, it gradually broke up – but its fragments were later sighted as far north as Latitude 36 degrees South – the latitude of Buenos Aires.

What is a tornado?

On Thursday, 14 December 1989, a tornado scythed through a small village, tearing off roofs and sending cars whirling like toys along the main street. This didn't happen in the American mid-west, a territory notorious for tornados. It happened in the Norfolk village of Long Stratton, between Norwich and Ipswich. The fierce wind destroyed a shop and damaged 100 houses, leaving some uninhabitable.

Tornados strike the British Isles more often than is generally realized. One of the worst recorded cases occurred on 8 Dec 1954, when a tornado struck a suburb of north-west London, destroying houses, factories and shops over a path 12 miles (19 km) long by

400 yd wide, seriously injuring 20 people. The wind was so fierce that it lifted a car nearly 20 ft (6 m) into the air in Acton and held it suspended for several seconds. And in August 1984, a tornado ripped through the village of Gotham, near Nottingham, causing £100,000 worth of damage in just a few seconds.

On average, tornados hit the British Isles on 30 days every year, with as many as 100 'twisters' occurring in a single day. In the swirling vortices of the worst ones, wind speeds can reach 175 mph (280 km/h) – more than twice the speed of hurricane-force winds.

So just what is this natural phenomenon that is capable of such destruction. In essence a tornado is a powerful whirlwind, formed when a rising current of warm air begins to rotate, creating a narrow vortex that is sustained by more currents of warm air which are dragged into it at ground level. Tornados are usually associated with cumulonimbus clouds and thunderstorms and may be set in motion by immense updraughts which rotate in the cloud itself. Reduced air pressure within the rotating vortex leads to the formation of water droplets in a swirling funnel that gradually descends to ground level, where it sucks up dirt and debris. If the phenomenon occurs over the sea, the result is called a waterspout.

Most tornados are created when thundery cold fronts sweep across the country from the west or north-west, although some are caused by isolated thunderstorms. These are usually of the supercell type, where warm, moist air enters the forward right flank of the storm at low levels. This causes a persistent updraught which is forced to turn as it ascends because of the variation or wind force with height (wind shear) and the proximity of a downdraught of drier, colder air, entering the storm from different directions. The initial updraught formation, which leads to the formation of a tornado, is anticlockwise in the northern hemisphere (and clockwise, in the southern hemisphere) over a large diameter (see Coriolis effect, p. 176).

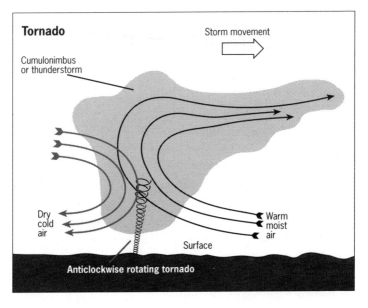

Tornado

Storm movement

Cumulonimbus or thunderstorm

Dry cold air

Warm moist air

Surface

Anticlockwise rotating tornado

The spiralling gradually extends upwards and downwards, with the speed of rotation increasing as the diameter decreases. The embryo tornado may remain hidden in the cloud for several minutes before a funnel begins to emerge from the cloudbase; when this reaches the ground, the result is a tornado. Occasionally, tornados can last for half an hour or more, but most range in duration from just a few seconds to three or four minutes.

Tornados have been known to produce some strange side effects, and account for some of nature's more bizarre mysteries. There have been occasions when frogs, fish, turtles and other creatures have fallen from the sky during rainstorms, with the only logical explanation being that they were sucked into the vortex of a waterspout and then released miles away when the 'twister' eventually dissipated.

What destroyed the walls of Jericho?

Scientists have been examining biblical and other ancient documents in an attempt to meet one of the most challenging needs of the 20th century, and in so doing they have solved one or two mysteries from the past.

Their aim is to find a foolproof way of predicting when an earthquake is likely to happen.

Unlike lightning, earthquakes tend to strike in the same place twice, and in Israel seismologists have been studying historical records to build up a dossier of earthquake activity extending over thousands of years. One of their aims is to establish whether there is a definite pattern to earthquake activity in much the same way that there appears to be a regular cycle of sunspot activity.

The Israeli scientists' research centres on the Jordan Rift, which has been subjected to earthquakes since the dawn of recorded history. The whole area is affected by the relentless northward movement of the plates of the Earth's crust that form Arabia. The Arabian Plate, in collision with Asia, is moving north at up to one centimetre a year. This does not sound a lot, but it is enough to produce ongoing seismological disturbances throughout the Middle East – particularly in the so-called Jericho Fault, a weak spot in the Earth's crust which has been hit by many earthquakes since the first one was recorded about 1000 BC. This was at the time of Joshua's famous siege, when the walls of Jericho collapsed. The event was accompanied by a massive mud slide that dammed the Jordan, allowing the Israelite army to pass over.

Scientists have recorded 32 earthquakes on the Jericho Fault since 117 BC, all of them well documented. The last one occurred in July 1927, when a strong earthquake caused cracks in buildings and fissures in the ground at Jericho and several other cities and villages in Judea, Samaria and Galilee.

In 31 BC, a major upheaval on the Jericho Fault destroyed the town of Qumran, where the Dead Sea Scrolls were found. A detailed description of this catastrophe was given by the Jewish historian Josephus: 'At this time it was that the fight happened at Actium between Octavius Caesar and Anthony in the seventh year of the reign of Herod and it was also that there was an earthquake in Judea, such a one as has not happened at any other time and which earthquake brought a great destruction to the cattle of that country. About 10,000 men also perished by the fall of houses, but the army which lodged in the field recorded no damage by this sad accident.' (Antiquities of the Jews, Book XV, chapter 5, verse 2.)

Isaeli experts think that an earthquake may also have been responsible for the destruction of Sodom and Gomorrah in about 2000 BC. Although the exact location of these ancient cities is not known, it is likely that they lay in the plain north of the Dead Sea, east of Jericho and very close to the line of the Jericho Fault. A big earthquake here would have produced massive clouds of dust from the dry ground, accounting for the biblical description: 'The smoke of the country went up as the smoke of a furnace.'

Early in 1992, a devastating earthquake at Erzincan in eastern Turkey produced a similar effect. The only difference was the comparison drawn by an eye-witness.

'A dust cloud rose above the whole city,' he said. 'It was like Hiroshima.'

What is ball lightning?

During the Second World War, the crews of high-flying bombers were sometimes mystified – not to mention somewhat frightened – to see glowing balls of light that looked almost solid appear without warning inside their aircraft, drift slowly down the centreline of the fuselage, and disappear out of the tail end.

No-one had an explanation for these apparitions, which the Americans inexplicably nicknamed 'foo fighters'.

In the post-war years, flying saucer enthusiasts latched on to the story and had a field day with it. The glowing balls, they claimed, were remotely controlled reconnaissance probes, despatched by alien mother ships in Earth's orbit to see what the natives were up to. This bizarre theory still continues to crop up in UFO literature – the wartime experience having been expanded by accounts of the objects appearing in civil airliners over the years – despite the fact that we now have a good idea of what the balls of light really are.

They are manifestations of one of the oddest phenomena known to science: ball lightning. This is a true curiosity of nature, and one that still defies a complete and satisfactory scientific explanation.

Ball lightning manifests itself in the form of glowing spheres that appear to move freely through the air, having the ability to penetrate solid objects and to appear inside buildings and aircraft. Usually, they range in size from about an inch to a few feet in diameter and are associated with thunderstorms, which is why they are linked with lightning – although they may have nothing to do with lightning at all. Ball lightning is normally yellow, red or white in colour, although it sometimes glows purple or green. Sometimes, too, the spheres emit sparks or rays. They materialize quite literally out of the blue. Mostly they glide through the air in complete silence, very infrequently producing a faint hissing noise. They can last for a minute or more before extinguishing, and their demise is almost invariably noisy; they explode or simply vanish with a loud bang, leaving a smell of ozone behind.

There is a lot of energy in a typical lightning ball. One estimate of the energy content was made when a ball obligingly popped into a barrel of rainwater and boiled it; it also struck a house, cut a telephone wire and burned a window frame. Calculations show that this particular lightning ball released several megajoules of energy, equivalent to the output from an electric heater with the power of one kilowatt for more than an hour.

Over the years, many theories have been put forward to account for ball lightning. In the 1970s, scientists at the National Center for Atmospheric Research in Boulder, Colorado, suggested that thunderstorms could behave like huge natural particle accelerators, producing protons with an energy as high as one megaelectronvolt. These protons could then induce nuclear reactions by colliding with atomic nuclei in the atmosphere, creating isotopes of oxygen and fluorine. The isotopes would then decay, emitting positrons and gamma rays which would provide the energy to create a lightning ball.

Another theory is that lightning balls are minute particles of antimatter. In theory, every fundamental type of particle has a corresponding antiparticle possessing the same mass but having an opposite charge. A key aspect of the theory is that when antimatter meets matter, there is a massive release of energy in the form of gamma rays, causing mutual annihilation.

The theory suggests that minute grains of antimatter filter down through the Earth's atmosphere, escaping immediate annihilation thanks to a potential barrier caused by certain quantum effects. The grains become negatively charged as they emit positrons and are drawn towards the ground during thunderstorms, eventually becoming unstable and exploding. It is probably the most plausible theory advanced so far, and one that might account for many other mysterious lights in the sky.

Of course, there really might be an alien mother ship out there somewhere, keeping an eye on us…

What keeps migratory birds on course?

Every autumn, a mass exodus gets under way as billions of birds set out across oceans and continents towards winter quarters that are often many thousands of miles from their summer nesting areas.

Most migratory birds – and that includes very many of the 8000 known and named species – nest north of the equator. This is not only because the climate is more comfortable, but also because there is 50 times more land area in the north than in the southern temperate zone, nine-tenths of which is covered by ocean. Nesting birds need a lot of territorial space, and the northern hemisphere, with land masses extending thousands of miles from the equator to the pole, has plenty of space to offer.

Some birds perform prodigious feats of endurance. The Arctic tern, for example, breeds in northern Canada, Greenland, Iceland and northern Europe but spends the winter months in southern Africa and Antarctica, which involves a round trip of some 22,000 miles (35,400 km) each year. Because it spends all its life living close to one or other of the Earth's poles, where the Sun never sets properly for weeks on end, it probably sees more daylight than any other living creature.

Just how some birds manage to stay on the wing for up to 30 hours remained a puzzle for a long time. However, quite recently, ornithologists discovered that before birds set out on their migratory flights they develop a layer of fat around the breast-bone, and this provides them with the stored energy they need for long-range flying.

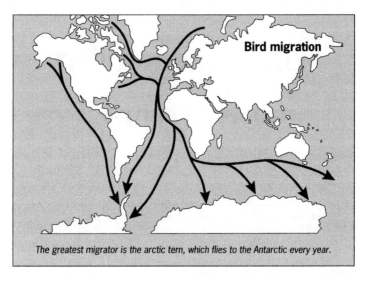

Bird migration

The greatest migrator is the arctic tern, which flies to the Antarctic every year.

It is only recently, too, that scientists have been able to plot the routes followed by migratory birds with any real accuracy. When they succeeded, they found that some routes were almost arrow-straight while others had curious twists and bends in them; sometimes, birds would even do a complete about-turn and fly back towards their departure point for a while before resuming their original course.

The scientists discovered it was all to do with the flight characteristics of individual species. Birds such as storks and cranes, which are basically gliders, need to use warm air currents as much as possible. Just like a glider pilot, a migrating stork will search for these currents (thermals) and will sometimes retrace part of its flight path to find them. Squadrons of storks will circle to great altitudes in the warm air that rises over an isthmus of land such as Gibraltar or the Bosphorus before setting out on a gliding flight over the sea.

Which brings us to the main question of how migrating birds maintain course once they are en route. The most generally accepted explanation is that birds, bees and other animals navigate by using the Earth's magnetic field and that they may actually be able to see the lines of force, possibly as stripes in the sky, which allows them to select their flight paths.

Scientists have found evidence that light is critical in the ability of some animals to navigate. Research indicates that such creatures may have special pigments and cells in their visual systems that allow them to perceive the Earth's magnetic fields as bright or shaded patterns. When the eye absorbs light, the pigment becomes weakly magnetic, and in so doing it alters certain nerve signals which the eyes send to the brain. The theory is that some brain cells contain crystals of magnetite, an iron oxide capable of detecting a magnetic field; migrating animals possibly use a crude 'visual' magnetic sense to detect compass directions and magnetite to detect local or global variations in the magnetic field.

The use of this dual navigational system might explain why some birds can navigate on dark and cloudy nights, as the light-sensitive pigment would work even in conditions of low illumination.

Scientists have also researched the possibility that the human brain may also contain crystals of magnetite, which opens up some interesting possibilities – and leads us neatly into the next topic.

What is dowsing?

Some time ago, researchers at the California Institute of Technology identified crystals of magnetite (FeO Fe2O3) – the iron oxide that helps migrating animals to navigate – in the human brain. Whether the crystals are sufficient to give humans an unconscious magnetic – and hence directional – sense is still under investigation, but they may provide the key to the ancient and rather esoteric art of dowsing.

The most commonly known form of dowsing is water divining, but dowsing embraces a whole spectrum of underground features including mineral veins. The dowser practises his art with the aid of a Y-shaped hazel twig which deflects strongly downwards at the source of the underground water or mineral; a metal rod or a pendulum may be used as an alternative. Many mining and prospecting industries use dowsing in mineral exploration; they call it the BPM technique, BPM being an abbreviation of Bio-Physical Method.

Critics claim that the success of dowsing depends more on a thorough knowledge of local geology than on the ability to detect magnetic anomalies in the environment. There is no reason, however, to dismiss it out of hand; after all such anomalies exist everywhere. Magnetic Anomaly Detection (MAD) has been used for years in tracking submerged submarines, whose metal mass causes fluctuations in the Earth's magnetic field. The theory that people can respond to magnetic cues provides a logical explanation of how dowsers are able to find veins of metal ore; in some cases the ore minerals are themselves magnetic. The presence of underground water can also produce large anomalies.

Russian geologists have made widespread use of dowsing since the

1970s, and have drilled thousands of test bores as a result of information gathered in this way. They report that the most frequent dowsing rod movements occur over deposits of sulphide or chromite, diamond pipes, fault and fracture zones, steel pipelines, and underground cavities. The rod also reacts strongly to overhead electricity cables.

It is becoming increasingly clear through recent scientific experiment that dowsing does depend on the existence of tiny magnetic receptors – magnetite crystals – in the human brain. Some of us seem to have a larger concentration of these crystals than others, which may explain why many people who attempt dowsing achieve no reaction at all.

Some scientists believe that people who are sensitive to magnetic anomalies could have as many as 85 million crystals of magnetite in their brain's tiny sensor, capable of detecting changes in magnetic fields as small as one nanotesia – less than one twenty-thousandth of the Earth's field. When this sensor detects a magnetic anomaly, it sends a signal to the dowser's muscles, which experience a small contraction causing the lightly held dowsing rod to deflect.

What is carbon dating?

Carbon dating – or radiocarbon dating, to give it a more correct title – was first developed by the American chemist Willard Libby in 1949. Radiocarbon, an unstable form of carbon, is found in all living things; plants absorb carbon dioxide from the atmosphere and incorporate it in their tissues, and some of the carbon dioxide contains carbon-14, the radioactive isotope of carbon. Animals absorb carbon-14 into their bodies as a result of eating plant tissues.

Radiocarbon begins to decay the moment it is formed. The proportion in the atmosphere, in CO2, is maintained at a constant level by new formation. Once fixed in organic matter, decay causes the radioactivity of the carbon-14 in the matter to become gradually less than that in the atmosphere. Carbon-14 in organic matter decays very slowly, having a half-life of 5730 years – in other words, half of it decays every 5730 years. So, if the carbon remaining in, say, a fossilized tree can be measured against the amount in a living tree, the age of the fossil can be determined.

The method yields reliable ages up to 50,000 years, but after 120,000 years so little carbon-14 remains that no measurement is possible. Willard Libby thought that the concentration of carbon-14 in the atmosphere had remained constant throughout the ages, but this assumption was later proved wrong. Radiocarbon datings from tree rings showed that the concentration was much higher before 1000 BC. This discovery led to the modification of the dating procedure; today, all radiocarbon dates are calibrated against calendar dates obtained from tree rings or against uranium/thorium dates from coral. The carbon-14 content is determined by counting beta particles with either a proportional gas or a liquid scintillation counter over a period of time.

The system, however, has a serious flaw. Since radiocarbon dating can detect only those carbon atoms currently in the process of decaying, often a substantial portion of a fossilized specimen has to be burned and its smoke collected and analysed in order to obtain a sampling. A new method, called accelerator mass spectrometry (AMS), is now in use; this system is over 1000 times more sensitive than radiocarbon dating, and needs only a tiny sample of a specimen to determine its age.

Admittedly, AMS is an extremely complex operation. First, carbon is chemically extracted from a sample. Next, bits of the element are placed in a Van De Graaf nuclear accelerator, a 150 ft (45 m) long, tubelike machine that bombards them with highly charged ions. This separates the carbon into its constituent atoms, electrically charging them in the process. The atoms are then attracted by an oppositely charged ball in the middle of the accelerator.

As the atoms pass through a small hole into the ball, their charges reverse. Since like charges repel one another, the ball ejects the now-similar atoms through another hole, like bullets from a machine gun. They are then sorted by electric and magnetic fields in the accelerator, after which nuclear detectors identify and count the radiocarbon atoms. This yields a direct measure of the amount remaining in the sample.

What is a supernova?

Let's be clear about one thing – in astronomical terms, a supernova is not related to a nova.

Novae (the Latin plural) are quite common. Every so often, a faint star – usually one of a pair in a binary system – suddenly flares into unexpected brightness, shining with up to 10,000 times its former intensity, and becomes a nova. Astronomers believe this happens when gas from one of the stars in the binary system flows to its companion star, most commonly a white dwarf – a small, hot star in the last stages of its life – and ignites. The explosion blasts the gas out into space at speeds of 930 miles/sec (1500 km/s) or more, but does not destroy the star. After a few weeks or months the latter subsides to its previous state, and in the course of time may become a nova several times over as the process is repeated.

A supernova, on the other hand, marks the violently explosive death of a star, which temporarily bursts into the brilliance of 100 million suns. Its brightness lasts for a few days or weeks, and then fades. On average, it is thought that a supernova occurs in a large galaxy once every 100 years, although some may go undetected because they are obscured by interstellar dust.

The build-up to a supernova begins when a heavy star, ten or more times as massive as our own Sun, has lived out the main sequence of its life. One such star, catalogued as Sanduleak –69 202, provided a typical example of the sequence that leads to total destruction.

Sanduleak was once a star in our neighbouring galaxy, the Large Magellanic Cloud, which is visible from the southern hemisphere. Sometime in the Old Stone Age, when man still sheltered in caves and was learning to work with rudimentary flint tools, Sanduleak exploded. About 170,000 years later, the news of the star's demise reached Earth in the form of a flood of neutrino particles and the light of the explosion. It was the first time that a supernova had been observed in the skies of our planet since 1604, and it gave astronomers an unparalleled opportunity to observe the awesome destructive forces that had torn the star apart.

Previous spectrum analysis had already shown that Sanduleak was a hot star, bluish in colour, with a surface temperature of 20,000 Kelvin – a temperature scale used by scientists, beginning at Absolute Zero, –459°F (–273°C), and increasing by the same degree intervals as the Celcius scale; 0 degrees C is the same as 273 K; the Sun's surface temperature is 5800 K. Sanduleak was a much younger star than the Sun – a mere 20 million years old against the Sun's age of 5000 million years – but it was 20 times the Sun's mass.

Sanduleak's temperature was 10 million K at its core, and this tremendous heat was its eventual undoing. The hydrogen nuclei at its heart began to fuse into helium. The reaction kept the star shining for most of its life, but it proceeded very rapidly because Sanduleak was so massive, and by about a million years ago all the hydrogen in its core had turned to helium.

The helium core shrank; the temperature rose to 50 million K; and nuclear reactions began to turn the helium into carbon. Simultaneously, the star's outer layers swelled until it became a red supergiant, several hundred times larger than the Sun. Its gravity was no longer powerful enough to hold on to its outer gases, which began to drift away into space. When it had lost about a quarter of its material in this way, it shrank to become a blue giant, about 50 times larger than the Sun.

By now the star's core had turned entirely to carbon, and as it shrank the temperature rose again. After a thousand years the carbon core had burned away to neon, and within a year this had in turn burned away to silicon. The temperature now rose to 400 million K, and the silicon at the core fused to create iron nuclei. Around this iron core were successive shells of silicon, neon, carbon and helium, enclosed by an outer shell consisting mainly of hydrogen.

One day, like the failure of a weak heart, the last nuclear reaction in Sanduleak's iron core flickered out and the star's centre collapsed catastrophically, releasing a flood of neutrinos. Sanduleak blew itself apart in a stupendous explosion, leaving only a cloud of radioactive debris, shining with the light of 250 million suns, to mark its passing.

What was the Star of Bethlehem?

Could the star that led the Magi from Chaldea to the stable in Bethlehem, and to the infant Jesus, have been a nova or supernova – or perhaps a comet or a planetary conjunction?

Theologians might accept that whatever celestial object pointed the wise men in the right direction was put there by divine intervention, but astronomers tend not to be so trusting in their belief, and the mystery has intrigued them for centuries.

Any attempt to solve the mystery, and to identify the celestial body that might have been visible at the time, is beset by problems, the biggest of which is in establishing the real date of the nativity. The only sure thing is that it was not 25 December; that marked the winter solstice and the Roman festival of Saturnalia, on which the feast of Christmas was superimposed over three centuries later, purely for reasons of convenience. Most of the major Christian feasts, in fact, replaced pagan ones. If the Gospel is to be believed, and shepherds were out in the fields looking after their flocks, it was probably during the springtime lambing season.

The year of the nativity is highly suspect, too. The established date is based on the researches of a monk called Dionysius Exiguus, who in AD 525 set out to confirm the time by a meticulous examination of events in Roman history that had unfolded since. However, he missed out four years of the rule of the emperor Octavian, and so if we want to establish the true date of Christ's birth we have to look several years beyond the accepted starting point of AD One.

So what confirmed and noteworthy celestial objects were around in 4 or 5 BC? The one that might have been favourite, Halley's Comet, is out of the time bracket since it appeared in 12 BC and did not put in another of its periodic appearances for 76 years. However, Chinese astronomers, who meticulously recorded everything of note that appeared in the sky, recorded a comet in 5 BC and observed it for 70 days before it faded from sight. Some astronomers think that they may actually have observed a nova or supernova, but this is unlikely; the Chinese sages were meticulous in their descriptions of celestial bodies, referring to comets as 'broom stars', because they appeared to sweep the skies, and to novae as 'guest stars' because they made only a brief appearance.

The remaining theory is that the ancient Middle Eastern astronomers might have witnessed a planetary conjunction – and in 7 BC, which is just within the acceptable time frame, there was a spectacular one, a triple conjunction of Jupiter and Saturn in the constellation of Pisces. In this event, which occurs only once in 139 years, Jupiter appeared to catch up with Saturn in May, and later in the year the two planets seemed to move in opposite directions as the Earth, on its inner orbit around the Sun, overtook them. In October, Jupiter passed Saturn going the other way, after which the two planets stayed close together until the end of the year. After a slight separation, the planets closed up again in December for a third conjunction before separating for another 139 years.

To the astronomers of Chaldea, who were also noted astrologers, this event would have been of considerable importance. So would a conjunction between Jupiter and Venus which, according to modern calculations, took place on 12 June in 2 BC.

But which brilliant object did appear at the nativity of Christ – and we have only the testimony of the Matthew Gospel that one appeared at all – remains a matter of conjecture. Perhaps it is fitting that this should be so; for the Star of Bethlehem will always be part of the magic of Christmas.

What is a pulsar?

When a star explodes, it leaves behind a remnant of its former self – a kind of cosmic corpse.

Back in the 1930s, two European astronomers working in the USA, Fritz Zwicky and Walter Baade, suggested that when a star becomes a supernova, its core collapses inwards to become a very compact star. The collapse packs the protons and electrons in the gas so tightly together that they combine to make neutrons. The resulting 'neutron star' may contain more mass than the Sun, but it is only a few miles in diameter and a pinhead of its matter might weigh a million tonnes.

For decades, this remained no more than a theory. Then, in the autumn of 1967, two radio astronomers at Cambridge, Tony Hewish and Jocelyn Bell, detected regular signals coming from the sky. The radio pulses were very regular and rapid, and came from a weak source which fluctuated rapidly, as though it were ticking.

Quickly dismissing any notion that the signals came from some alien civilization trying to establish contact with us, the astronomers named the object a pulsar. The announcement of its discovery was officially made on 29 Feb 1968, and it was designated PSR (Pulsating source of Radio) 1919. What Hewish and Bell had found was a kind of cosmic lighthouse, an object spinning at a fantastically rapid rate – about once a second – and emitting beams of radio waves as it did so. They had, in fact, discovered a neutron star.

The star lay in the constellation Vulpecula. Since 1967 hundreds more have been located. They are dense objects only about 12.5 miles (20 km) across. Their rate of spin varies; PSR 1919, for example, sends a radio beam in the direction of Earth every 1.3 seconds, which is the time it takes to spin. The density of a neutron star is about 100 million million times that of water; a teaspoonful of its material would weigh 100 million tonnes, and if you were able to stand on one, its gravity is such that you would weigh 10,000 times more than you do on Earth.

Some pulsars, all of which are thought to be rapidly rotating neutron stars, lie in the gaseous remnants of the supernovae that gave birth to them. PSR 0531+21, for example, lies in the Crab Nebula, the remains of a supernova that exploded in 1054 in the constellation Taurus, which was logged by Chinese astronomers. The youngest

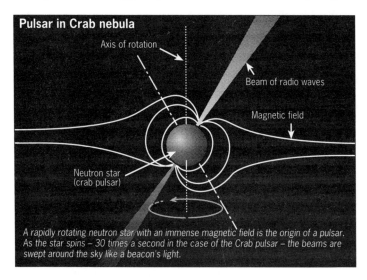

Pulsar in Crab nebula

Axis of rotation

Beam of radio waves

Magnetic field

Neutron star
(crab pulsar)

A rapidly rotating neutron star with an immense magnetic field is the origin of a pulsar. As the star spins – 30 times a second in the case of the Crab pulsar – the beams are swept around the sky like a beacon's light.

pulsars, like this one, spin the fastest – a period of just 0.03 seconds in this case – and spin more slowly as they grow older, until after about ten million years the rotation becomes so slow that the pulsar no longer gives off radiation.

In 1982, astronomers got a surprise when they discovered a pulsar that spins even faster than the Crab Nebula pulsar – 20 times faster, to be precise. Like the first neutron star ever identified, this one, named PSR 1937+21, lies in Vulpecula and spins at the astonishing rate of 1.6 milliseconds, so astronomers called it a 'millisecond' pulsar. More such pulsars have been discovered since, and the surprising thing about them is that they are not young, as might be expected; all of them are more than a billion years old – as much as 15 billion years old, in some cases.

There may be more surprises in store, because in 1991 astronomers at Jodrell Bank Observatory in Cheshire thought they might have detected a planet orbiting the pulsar PSR 1829-10, in the constellation Scutum. It isn't an impossibility; even if the Sun became a supernova, Earth would survive the explosion. It's just that we would not.

What is a black hole?

Sometimes, if the collapsing core of a star is too massive – heavier than three suns – it cannot become a neutron star. Its gravity is so powerful that the core continues to collapse until the matter that formed it is crushed completely out of existence. The core then becomes a mathematical point, with no size at all and an infinite density. Surrounding this point is a gravitational field so strong that nothing can escape from it – not even light. Its pull of gravity would be about 1600 million billion times the force we experience on Earth. So a black hole is created; it is a hole because anything that falls into it can never emerge again, and it is black because it does not let light escape. Even if you tried to illuminate it – assuming you could get close enough – the hole would swallow up the beam from your torch.

The evidence for the existence of black holes has emerged gradually over the years. As with neutron stars, astronomers first predicted black holes in the 1930s, but proving their existence was a difficult task, because they can only be detected by observing their effect on other objects.

The first positive indication that black holes did exist came when astronomers observed a star in the constellation Cygnus in 1971. Named Cyg X-1, it orbits around a companion star that is invisible through ordinary telescopes. By careful observation of the visible star, astronomers found that the invisible companion exerted the gravitational pull of an object as heavy as ten suns. It was much too heavy to be a neutron star, so the only possible conclusion was that it was a black hole.

Apart from Cyg X-1, only three other possible black holes had been detected in our galaxy by 1995. Some were found because they appeared to be sources of X-ray emissions. Why, you may justifiably ask, do X-rays escape from black holes when light does not? The answer is that the X-rays come from hot gases that lie just outside the invisible barrier – the so-called event horizon – at which escape velocity just equals the speed of light. As gas falls into a black hole, its angular momentum is preserved: the gas forms into a disk, spiralling faster and faster as it approaches the hole's centre of mass. Atoms smash into one another with increasing frequency and force, and friction heats the gas to 100 million°C. X-rays are the by-products of these violent atomic collisions.

Where the gas comes from depends on the size of the black hole. When a star at least 3.2 times as massive as the Sun collapses in a double star system, the black hole can still draw gaseous material from the outer regions of its still-burning companion star.

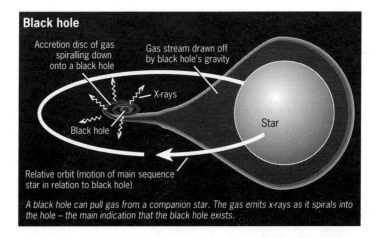

Black hole

Accretion disc of gas spiralling down onto a black hole

Gas stream drawn off by black hole's gravity

X-rays

Black hole

Star

Relative orbit (motion of main sequence star in relation to black hole)

A black hole can pull gas from a companion star. The gas emits x-rays as it spirals into the hole – the main indication that the black hole exists.

There is evidence to suggest that black holes with a mass several million times that of the Sun lie at the centres of galaxies (including, some astro-physicists suspect, our own Milky Way galaxy) and globular star clusters. The density of stars in these regions is 1000 times what it is in our own part of the galaxy, so that individual black holes will tend to collide with one another, growing inexorably bigger over billions of years. The gravity of a super-massive black hole is strong enough to pull in the diffuse gases from between the stars and compress it into a hot disk that emits X-rays.

As we said earlier, matter and energy appear to be crushed completely out of existence inside a black hole. Yet this idea defies logic, for nothing in the universe is completely destroyed; it changes into something else. And that brings us to one of the most intriguing questions of modern science.

What is time – and can we travel through it?

Where does the material go that is sucked into a black hole? According to some theories, it is instantaneously hurled as a spray of energy into a different part of the universe – perhaps even into a different time. The complex laws of Quantum theory allow a particle, such an electron, to tunnel through an energy barrier like that surrounding a black hole. In theory, it then enters a different region of space and time – another universe.

When he was developing his theory of relativity, Albert Einstein showed that in many respects time was like an extra dimension to space; so space and time may be considered as entwining into a single entity, rather than two separate things. This single entity, called space-time, is therefore considered to have four dimensions: three of space and one of time.

How we fit into this concept requires some explanation. We exist in three dimensions; our bodies occupy a certain volume, and we can move them around from place to place. We also know that there is a fourth dimension – time – because it tells us when we are in a particular place, where we were in the past, and where we may be in the future. So to tell where we were, are, or will be needs a combination of all four dimensions, which are inseparable.

We are used to thinking of time as a dimension that moves forwards – the so-called 'arrow of time' – a concept that is probably generated by the fact that we are born, grow up, live and die. It's the same in relation to the universe, which according to one widely held theory is born in an explosion, expands to a certain limit, then starts to contract until the process starts all over again.

In simple terms, theories about time travel – at least until now – have been based on the possibility of travelling faster than light. The electromagnetic radiation we call light travels at 186,000 miles/sec (299,300 km/s), and how we see objects depends on how long the light from them takes to reach us. In other words, everything we see in the universe is in the past.

Einstein's theories say that to accelerate any material object up to and beyond the speed of light would take an infinite amount of energy and would therefore be impossible. What they do not say is that nothing can travel faster than light.

We have measured the speed of light and know what it is. It has a definite boundary, and so, scientists argue, there must be something on the other side – a universe we can neither see nor touch. They have even given a name to the particles inhabiting this imaginary universe, calling them tachyons. These particles, the theory goes, have always travelled at faster-than-light speeds, so infinite energy would be needed to slow them down to the speed of light. The theory also supposes that they must be travelling backwards through time, from the end of the universe towards its birth – time's arrow in reverse, as it were.

It looks simple, really. If tachyons exist, so does the reality of time travel. Many scientists dismiss such a notion as nonsense, claiming that time travel goes against all the laws of physics and that we are trapped in the little block of space we call the present, like it or not. However, in 1991, Dr Amos Ori of the California Institute of Technology suggested that quantum effects might make it possible to 'tunnel' through space near a black hole and emerge in another region of space and time. This idea was also dismissed on the grounds that anything approaching the immense gravity of a black hole would be warped out of existence.

However, the theory has been advanced by Ori and other physicists that an object can approach the boundary, or event horizon, of a black hole as closely as it likes without being destroyed. Instead, the hole's massive gravity might be used to accelerate a spacecraft to fantastic speeds and catapult it on a journey to the stars – perhaps even use the loops of warped space-time that girdle a black hole's boundary to hurl itself back into the past, or even into a parallel universe.

Fantastic? Maybe. Only time will tell.

What was the Big Bang?

Before the beginning, it is believed there was nothing; only a vacuum, and 'perfect symmetry'; in other words, an unchanging state. Then, out of nothing, the universe erupted, born in a searing, unimaginable fireball. At that precise instant everything – all matter, energy, the very fabric of space and time – came into being. This was the Big Bang.

An instant later – that is to say, one million-million-million-million-million-millionth of a second later – the universe was the size of a pea. It was also a very hot pea, with a temperature of 10,000 million million million million°C. As the pea began to expand, however, it also began to cool, so that a second after the Big Bang the temperature had fallen to 10,000 million°C. It has been falling ever since.

As the cooling process continued, the fundamental particles created in the furnace of the primordial explosion that occurred 15,000 million years ago – electrons and positrons – annihilated each other to form photons, although a remnant of electrons was left over. All this happened about half an hour after the explosion.

Some 300,000 years after the Big Bang, the temperature of the expanding universe had dropped to 5400°F (3000°C), about the same as the temperature of the Sun's surface, and this was cool enough for atoms to form. Nuclei of hydrogen and helium mopped up the free electrons, and the photons, free of their influence, scattered to the limits of the universe. They are still there today, after 15 billion years, forming background radiation that can be detected as evidence of the whole stupendous event.

The departure of the photons meant that particles of matter, so far kept apart by the radiation, could now coagulate by gravitation to form galaxies, which in turn fragmented to form stars. Our own Milky Way galaxy, 10,000 light years in diameter, grew to contain at least 100 billion stars, from which our own fairly insignificant Sun formed about 4.7 billion years ago. The Earth itself is just over 4 billion years old.

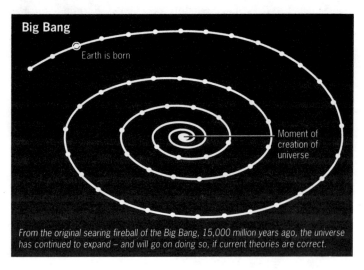

Big Bang

Earth is born

Moment of creation of universe

From the original searing fireball of the Big Bang, 15,000 million years ago, the universe has continued to expand – and will go on doing so, if current theories are correct.

The universe continues to expand, but one day, if it contains sufficient mass, the combined gravity of all the galaxies will bring the expansion to a halt and reverse it, causing a dramatic collapse that will end in a 'big crunch'. On the other hand, if there is insufficient mass to halt expansion, the universe may continue to expand until all the stars of all the galaxies eventually burn themselves out. Even if the collapse does take place, it may be followed immediately by another Big Bang.

Much evidence in support of the Big Bang theory is being obtained by the Hubble Space Telescope, named after Edwin Hubble, the American astronomer who in the 1920s discovered that the galaxies were receding and that the universe was therefore expanding. The concept is now generally accepted in place of the Steady State theory, which suggested that the universe had no origin and the matter in it was being continuously created.

But no-one, as yet, has come up with a satisfactory explanation as to what lit the fuse that touched off the Big Bang in the first place...

What is the Milky Way?

If you look up on a clear, moonless night, and you are well away from sources of light pollution, you will see a faint band of misty light stretching from north-east to south-west running from horizon to horizon. The best time to see it is in August, particularly at midnight, when it is almost directly overhead. It passes through many well-known constellations: Auriga, Perseus, Cassiopeia, Cygnus, and down towards Sagittarius, in the southern sky. In the southern hemisphere, observers can see its broadest and most luminous parts as it stretches from south to north through Crux (the Southern Cross), Centaurus and Sagittarius to Aquila and beyond.

The Milky Way was a marvel and a mystery to humankind from the dawn of recorded history until the early part of the 17th century, when Galileo turned his telescope upon it and found that it consisted of the combined light of millions of stars. However, it was another century and a half before anyone had a real inkling as to what it might be. In 1750 the Englishman Thomas Wright, a clockmaker's apprentice from County Durham who later became a mathematician and astronomer, suggested that the appearance of the Milky Way was due to the fact that the stars were arranged in a vast, flattened disk. Thus, anyone looking along the plane of the disk towards its centre sees innumerable stars, but comparatively few when looking the other way, towards its outer limit.

Further observations by William Herschel in the late 18th and early 19th centuries took the concept a step further. Herschel, who

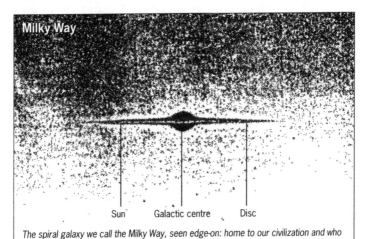

Sun   Galactic centre   Disc

The spiral galaxy we call the Milky Way, seen edge-on: home to our civilization and who knows how many countless others?

was attempting to map the distribution of stars in space, concluded that the Sun lay at or near the centre of the Milky Way, and it was not until 1917 that this idea was dismissed by the American astronomer Harlow Shapley. While surveying the distribution of globular clusters – massive clusters containing tens or even hundreds of thousands of stars – he found by measuring their distances and directions that they were grouped in a huge sphere centred on a point in Sagittarius. His conclusion was that the centre of the Milky Way disk had to coincide fairly closely with the distribution of the clusters, and that the Sun was in fact a long way from the centre.

We know now that the Milky Way is a spiral galaxy, of which the Sun and its planets form a tiny part. It is flattened in shape, with a central bulge of old stars surrounded by a disk of younger stars, arranged in spiral arms like a Catherine wheel. The whole system contains about 100 billion stars and is nearly 100,000 light years in diameter, with a thickness of some 2000 light years. Somewhere in the the central bulge, which is between 15 and 20 thousand light years across, is the galactic nucleus. Our Sun with its solar system lies almost 30,000 light years from the nucleus, about three-fifths of the distance from the centre to the outer rim, and follows a virtually circular orbit around the nucleus at a speed of 562,000 mph (900,000 km/h.) The solar system takes 220 million years to complete a full orbit; the last time it was at this spot, reptiles dominated the Earth and the mammals that were our remote ancestors were just beginning to evolve.

The Milky Way is not a highly concentrated galaxy. Much of it is empty space, the stars spread as thinly as a handful of sand grains scattered across the British Isles. Large tracts of it are also obscured by interstellar clouds of dust and gas, mainly hydrogen; as much as 90 per cent of the galaxy's mass may be composed of invisible dark matter.

The Milky Way is part of a local group of galaxies, the largest of which is M.31, the Andromeda spiral. This is one and a half times the size of our galaxy and is 2.2 million light years away from Earth. The closest galaxies are the Large and Small Magellanic Clouds, at 160 and 190 thousand light years respectively. The Large Magellanic Cloud orbits the Milky Way galaxy, trailing a long stream of hydrogen gas that is drawn off by the latter's immense gravity. Globular clusters lie around the periphery of the Milky Way galactic system, all of them very remote from us.

What is the zodiac?

The name Zodiac comes from the Greek word *zodiakos*, meaning 'to do with animals'. The plural *zodia* was used by Aristotle to describe the constellations, many of which were pictured in animal form by the ancient Greeks and other early civilizations.

The Zodiac itself is a faint band of fixed stars across the sky through which the Sun, Moon and planets travel. Because these objects lie near the ecliptic – the plane of the Earth's orbit and the path that the Sun appears to follow through the seasons of the year – the zodiac is quite narrow, extending not more than nine degrees on either side of the ecliptic.

Knowledge of the zodiac probably extended back to the Babylonians, who were the first to group stars into constellations over 3000 years ago, but it was the Greeks of the second century BC who organized the zodiac mathematically by calculating its 12 segments, or signs, as 30 degrees of arc each. The 12 arcs corresponded to 12 constellations, and since most were represented by animals, the Greeks called the band across the sky *zodiacos kyklos*, or 'circle of animals'.

Exactly when the zodiac was divided into the 12 signs is not known for certain, but it was doubtless long after men had identified the planets and endowed them with divine significance, probably for no other reason than that they always saw the same celestial bodies traversing the same band of sky. At first, no distinction was made between the constellations and signs of the zodiac; but then they started to take the equinoxes into account.

The equinoxes are the points in spring and autumn at which the ecliptic, the Sun's path, crosses the celestial equator, so that day and night are approximately of equal length. The vernal equinox occurs around 21 March and the autumnal equinox around 23 September. When the Sun crosses the vernal equinox, marking the beginning of spring, it enters both the constellation and the sign of Aries. But the vernal equinox (the point in which the Sun in its apparent annual motion crosses the plane of the Earth's equator from south to north) is not a fixed point in the sky. It moves westward, completing a full circle every 26,000 years. In astronomy, this is known as the precession of the equinoxes.

Over the past 2000 years, the vernal equinox has precessed through almost 30 degrees of arc, so that today, spring begins when the Sun is in the constellation of Pisces and the sign of Aries. This disparity makes no difference to modern astrologers, who deal with 12 exactly equal sections of the total circle of the zodiac, each one measuring 30 degrees, irrespective of what stars are contained in it.

The fact that there are only 12 signs of the zodiac is something of an anomaly; since the Sun traverses 13 constellations in its path across the sky, it would be logical to assume that there would be 13 zodiacal signs. The astrologers of ancient Egypt, Assyria and Chaldea recognized this problem and some of their zodiacs incorporated the 13th sign of the Serpent as a separate entity.

The Serpent coincides with the constellation Ophiuchus and lies between Scorpio and Sagittarius. In mythology, Ophiuchus is identified with the son of Apollo and Corinis. He became so skilled in medicine that he not only healed the sick, but also brought the dead back to life. This alarmed Pluto, ruler of the Underworld, who feared that his kingdom might become depopulated as a result of Ophiuchus's powers. Pluto made out a strong case to Jupiter, the god of gods, who had Ophiuchus killed with a thunderbolt. Afterwards, Jupiter was sorry for what he had done and had the healer elevated to a permanent place in the sky.

Today, there is a move to restore the Serpent to the zodiac. Sometimes, it is represented by a man with a fish's tail, a serpent twined around his lower part; other ancient zodiacs portrayed a serpent with wings or two heads.

What are the asteroids?

By the end of the 16th century, astronomers had already noticed that there was a gap of considerable size between Mars and Jupiter, and were speculating that the gap might harbour a missing planet. There seemed to be a definite geometrical progression in the distances of Mercury, Venus, Earth, Mars, Jupiter and Saturn from the Sun, which might be summed up by taking the numbers 0, 3, 6 12, 24, 48, 96, 192 and 384 – each one, apart from 0, being double its predecessor. By adding 4 to each and taking the mean distance of the Earth from the Sun as 10, the remaining figures gave the mean distances of the planets with considerable accuracy as far out as Saturn, which was the outermost planet known when the German astronomer Johann Bode published these findings in 1772.

In 1781, the English astronomer William Herschel discovered a seventh planet – Uranus – right where Bode's Law had predicted it would be, and his find caused a furore in the scientific world of his day. If Bode's Law was correct, as it now appeared to be, then there ought to be a planet between Mars and Jupiter – but there was no sign of it.

For 20 years, astronomers scanned the skies in an attempt to locate it and carry off one of the considerable prizes being offered to the first man to make the discovery, but no really systematic search was carried out until 1800, when a group of six astronomers styling themselves the 'Celestial Police' met at the observatory of Johann Schröter in Lilienthal, near Bremen and began to hunt the theoretical planet that corresponded to Bode's number 28.

On 1 Jan 1801, it seemed that the Italian astronomer Giuseppe Piazzi, of Palermo, had found the missing planet. Piazzi was not a member of the Lilienthal six, but he was to join them later. He named the newly discovered object Ceres.

Closer investigation, however, showed that it was far too small to be a planet, and its orbit was more elliptical and tilted than those of the planets that were already known. Over the next few years three more similar objects, Pallas, Juno and Vesta, were also found, and astronomers began to speculate that there had indeed once been a planet between Mars and Jupiter, but that some cataclysm had caused it to break apart.

The objects became known as asteroids, or minor planets. The search continued, but it was decades before another was discovered; this was Astraea, in December 1845. Since then, thousands more have been located; to date (1995) astronomers have calculated precise orbits for around 6000, and hundreds more are found every year.

Originally, asteroids were named after gods and then goddesses of Greek mythology, and were also given a number indicating their order of naming, so that 1 Ceres is the first one ever discovered and named.

So many have now been discovered, though, that they are named after pretty well anything, from 2001 Einstein to 2309 Mr Spock.

Not all asteroids are restricted to an orbit between Mars and Jupiter. Some have orbits that take them as far out as Saturn, and others come within the orbit of Mercury, the closest planet to the Sun. About 70 approach the orbit of the Earth, and are known as near-Earth asteroids. The main-belt asteroids – the tens of thousands orbiting between Mars and Jupiter and comprising about 95 per cent of all asteroids – are of enormous interest to astronomers, because they are probably related to material that formed as the solar system condensed from its parent nebula of cloud and dust. Some of the most distant asteroids, the Trojans, share the orbit of Jupiter in two clusters, one preceding the giant planet around the Sun by 60 degrees, the other following it by 60 degrees.

It was not until recently that we were able to take a close look at an asteroid. It happened on 29 Oct 1991, when the spacecraft Galileo, bound for Jupiter, passed within 1000 miles (1600 km) of the asteroid Gaspra and transmitted photographs of it back to Earth, revealing a potato-shaped lump of rock, possibly a fragment of a larger asteroid that had shattered in a collision. In 1993 Galileo also photographed asteroid Ida, which proved to have a small fragment in orbit around it, like a tiny moon.

What are shooting stars?

Every August, skywatchers are treated to one of the year's best celestial firework displays, caused when a shower of meteors, the Perseids – so called because they appear to come from the direction of the constellation Perseus – encounter the Earth. They are the brightest and most prolific of all meteor swarms, and for three weeks or so they leave their calling cards in the form of glowing streaks across the sky – streaks that give meteors their popular name of shooting stars.

Swarms of meteors orbit the Sun as do the planets, asteroids and comets. They are all named after constellations. Apart from the Perseids, there are the Lyrids, which appear in April, the Aquarids in May, the Orionids and Taurids in October, the Leonids in November and the Geminids in December. One swarm, which appears for only a short time in January, is named after a constellation that is no longer recognized, except on very old star maps: Quadrans Muralis, which gives the meteors their name of Quarantids.

Every day, over a million meteors visible only with astronomical instruments, and 500,000 more visible with the naked eye burn up in the Earth's atmosphere; their combined total weight is in the order of 5 tonnes, although only a fraction of this reaches the Earth's surface. Meteors that are massive enough to penetrate the atmosphere and impact on the surface are called meteorites; the largest known was discovered at Hoba, South-West Africa (later Namibia) in 1920 and is a block of iron measuring 9 × 8 ft (3 × 2.5 m) and weighing an estimated 60 tonnes. It fell to Earth in prehistoric times. The biggest and most famous meteorite crater lies in Arizona, near Diablo Canyon. About 25,000 years ago, a huge mass of iron and nickel, more than 200 ft (60 m) across, hurtled down from space and impacted on the Earth's surface with an explosive force of 30 million tonnes of TNT (30 megatons). It gouged a crater over 4000 ft (1200 m) in diameter and 600 ft (180 m) deep.

Fortunately, there have been only a few such occurrences throughout the Earth's history. The meteors that come to us in annual showers are harmless. Meteor swarms travel through space on parallel paths around the Sun, and the Earth meets them at different times of the year at various points in its orbit. When a meteor, travelling at up to 45,000 mph (72,000 km/h) enters the upper atmosphere, the friction sets up such intense heat that the particle is destroyed almost instantly in a glowing streak of radiation. The next time you see one arrowing across the sky, remember that you are watching something that is happening between 70 and 100 miles (112 and 160 km) high, and you will have some idea of the particle's fantastic speed.

Nowadays, scientists generally believe that meteors are the dust-like remnants of asteroids and comets, the product perhaps of collisions in deep space. Some meteors have been found to contain organic substances, and research has been in progress for some time to establish whether these were formed as a result of living processes, or by chemical reactions. There is even a theory that the first spores of life to develop on Earth were brought by a meteor.

It has been calculated that, on average, three or four meteors weighing 10 lb (4.5 kg) strike the Earth's atmosphere daily, and that one weighing 5 tonnes arrives once a month, although only about 1000 lb (450 kg) reaches the surface. Because of the large areas of ocean that cover the planet, most of these impacts pass unnoticed. A 50-tonne meteor arrives every 30 years, a 250-tonne object every 150 years, and a 50,000-tonne monster every 100,000 years – quite big enough to give planet Earth a real headache.

What are sunspots?

As long ago as 28 BC, keen-eyed Chinese astronomers, who meticulously logged everything unusual and noteworthy that appeared in the skies, recorded the presence from time to time of dark blemishes on the face of the Sun. Early telescopic observations in 1610-11 confirmed that such spots did indeed exist, although their nature remained a mystery.

Regular observation since those early days has taught astronomers a great deal about sunspots, which are visual evidence of continual activity in the photosphere, the visible surface of the Sun. They can range in diameter from about 600 miles (960 km) in the case of individual spots, to more than 60,000 miles (96,000 km) in the case of a group. The largest sunspot group recorded to date, in April 1947, had a diameter of 81,250 miles (130,700 km).

A sunspot is actually an area of gas that is cooler than the surrounding photosphere, and is thought to be caused by strong magnetic fields that block the outward flow of heat to the Sun's surface. It consists of a dark central umbra, a region of shadow that is totally dark and has a temperature of 4000 K (6700°F/3700°C) and a lighter surrounding penumbra of about 5500 K (9400°F/5200°C).

The photospheric magnetic fields that create sunspots are very powerful, usually ranging from 2000 to 4000 G – about 10,000 times stronger than the field at the Earth's surface, which varies from 0.3 G at the equator to 0.7 G at the poles. The centre of activity that is a sunspot develops in several stages, starting with the formation of a bipolar magnetic field that is accompanied by the sudden appearance of a small bright spot called a *facula*. *Faculae* (the Latin word meaning 'torches') were discovered early in the 17th century by the

German Jesuit astronomer Christoph Scheiner, and are now known to be luminous gas clouds, composed mainly of hydrogen, lying just above the photosphere.

The *facula* and associated magnetic field continue to develop, and after several hours one or more dark spots appear at the centre. This phase may last for several weeks. In the next stage, the spots disappear, the brightness of the *facula* decreases and long filaments begin to form. These in fact are solar prominences – streams of gas projected from the Sun into space – seen against the brightness of the solar disk. In the last stage, when there is no longer any visual evidence of sunspot activity, the magnetic fields continue to exist for some time.

Observation of sunspots over a period of time has revealed that the Sun follows a cyclic pattern of activity. Over a period of three or four years, sunspot activity increases to a maximum, then declines over a period of five to seven years. At this stage all spots have practically disappeared, and there is a period of minimal solar activity before the cycle begins again, the whole process taking 11.3 years.

Sunspot activity emits enormous amounts of electromagnetic energy, some of which reaches the Earth as charged particles, and for years scientists have been trying to verify a long-suspected link between this activity and fluctuations in the Earth's climate. A strong clue that such a link does exist is provided by a radioactive isotope called beryllium-10, which is created in the atmosphere when cosmic radiation interacts with atmospheric elements. The radioactive particles are eventually captured by raindrops and carried to earth, where they rest deep in the planet's crust.

Beryllium-10 is very easily created, and only a big upsurge in sunspot activity can slow down its production. When solar radiation reaches the Earth's atmosphere, it distorts and weakens normal cosmic radiation, with the result that less beryllium-10 is produced.

To find out how beryllium-10 levels have changed over the centuries, scientists took core samples from Green Lake, New York, where the 200 ft (60 m) deep sediment had lain undisturbed for about 10,000 years. In a layer of sediment deposited during the 17th century, they found a dramatic increase in beryllium-10. This meant that there had been a dramatic decline in sunspot activity, confirming the findings of the British astronomer E.W. Maunder, who in 1890 found that there had been virtually no spots at all between 1645 and 1715. In the years that followed, the northern hemisphere suffered an unusually cold spell known as the Little Ice Age.

What is the solar wind?

Every second, the Sun discharges more than a million tonnes of material into space in the form of electrically charged particles, mainly protons and electrons. Travelling at between 200 and 600 miles/sec (300 and 1000 km/s) it streams to the outer reaches of the solar system and beyond. This is the solar wind.

Its origin lies in the corona, the faint halo of extremely hot and tenuous gas boiling from the Sun's surface. Within the corona plasma is created, an ionized gas which contains positive and negative charges in approximately equal numbers. It is produced at a temperature of about a million degrees C; it is this plasma that constitutes the solar wind and permeates the vast region of space around the Sun known as the heliosphere.

Tenuous though it may be, the solar wind exerts a considerable force on the solar system. Its action on the dusty material that is shed by comets as they approach the warmth of the Sun ensures that their luminous tails always point outwards, and it is also responsible for disturbances in the Earth's magnetic field, the most obvious visual manifestations of which are the aurorae.

The theory that streams of particles emanating from the Sun have an effect on the Earth's magnetic field was first put forward in 1931, and in the 1950s observations of the ionized parts of comets' tails indicated that the solar wind was always present; this was later confirmed by satellite observations.

Although the solar wind is continuous, it is not constant in velocity or density. Every so often, the Sun discharges a massive bubble of plasma in what is known as a coronal mass ejection, the process taking about two hours. One such burst was recorded by the Giotto spacecraft on the way to its rendezvous with Halley's Comet in 1986.

The speed of the solar wind, driven outwards from the Sun by the high pressures within the corona, may be as high as 560 miles/ sec (900 km/s), although the average velocity is 250-312 miles/sec (400-500 km/s). The particles take four or five days to reach Earth.

The particles that comprise the solar wind are ejected as the Sun rotates, rather like water droplets from a garden sprinkler. Close to the Sun particles are trapped by its magnetic field and co-rotate with it, but past a certain distance the energy of the plasma exceeds that of the field and the particles move outward from the Sun in a spiral pattern.

The Earth's magnetic field acts as an obstacle to the solar wind; the plasma follows the field lines and is trapped around the planet in a region known as the magnetosphere, which is bounded by the magnetopause and stretches out in a long tail downwind of the planet. Because of this, only a fraction of the energy flowing through interplanetary space is absorbed by the Earth. It is nevertheless significant; the solar wind delivers one and a half times as much power as the human race is currently using from all sources, even though this is still 10,000 times less energy than sunlight supplies to the Earth.

What is light?

Of all the natural phenomena that surround us, light must surely be the one we take most for granted. Yet its nature has been the subject of scientific study since the days of ancient Greece. We know now that it is composed of electromagnetic waves, and that at the root of it all is a particle of the electromagnetic field called a photon. It is in fact the elementary particle of energy in which light and other forms of electromagnetic radiation are emitted. It has both particle and wave properties; it has no electric charge and is considered to be without mass, although it possesses momentum and energy. It is classified as a gauge boson, which is an elementary particle that carries the four fundamental forces of the universe (gravity, electromagnetic, weak nuclear and strong nuclear). As such, it cannot be subdivided.

In bright sunlight, a million million photons fall on an area the size of a pinhead every second. In quantum theory, the basis of particle physics, the energy of a photon is represented by the formula $E=hf$, where h is Planck's constant and f is the frequency of the emitted radiation. Planck's constant, named after the celebrated German physicist Max Planck (1858-1947) is a universal constant that is the energy of one quantum of electromagnetic radiation (the smallest possible package of energy) divided by the frequency of its radiation.

In a vacuum, the speed of light (and for that matter all other forms of electromagnetic radiation) is 186,000 miles/sec (300,000 km/s). In one year, therefore, the photons that compose a beam of light travel about 5.88 trillion (million million) miles (9.46 trillion km). This is expressed as one light year, a convenient unit for the measurement of interstellar distances. For example, Proxima

Centauri, the nearest star to the Sun, is 4.2 light years distant, so a round trip there and back at the speed of light, with a very quick stopover, would take eight and a half years. The fact that light had a finite speed was first established by the Danish astronomer Ole Romer (1644-1710), who reached an accurate conclusion by timing the eclipses of Jupiter's satellites.

Light is a visual manifestation of the electromagnetic spectrum, and sunlight, or white light, is composed of a mixture of light of different colours that can be separated into its components by dispersion, such as when each component wavelength is bent to a slightly different extent by passage through a prism. This fact was discovered by Isaac Newton in 1672; he used a prism to separate sunlight into its component colours of red, orange, yellow, green, blue, indigo and violet – the colours of the rainbow, dispersed by the prismatic effect of moisture droplets (Richard Of York Gave Battle In Vain is still the best way to remember the order). The wavelength of white light ranges from about 400 nanometres in the extreme violet to 770 nanometres in the extreme red, a nanometre being a billionth part of a metre.

What is a light sail?

Ever since men began to dream of the possibility of interstellar travel, astrophysicists have been preoccupied with the main obstacle in the path of realising this goal: crossing the vast gulf of space that lies between the stars without using up several lifetimes. To give an example: the space probe Voyager 2, travelling at 36,250 mph (58,000 km/h) will take over 80,000 years to cover 4.3 light years, which is the distance between our solar system and the nearest star, Proxima Centauri.

In the 1920s, the visionary Russian scientist Konstantin Tsiolkovsky, rightly described as the father of astronautics, postulated the idea of using the direct pressure of sunlight to drive a spacecraft. He and his fellow scientist, Fridrich Tsander, referred in their notes to using 'tremendous mirrors of very thin sheets' and to 'using the pressure of sunlight to attain cosmic velocities'. But it was not until 1958 that interest in the idea was reawakened by Richard Garwin, a consultant to the US Department of Defense, who published a paper on the concept of what was by then termed solar sailing.

The concept was taken a stage further by the Anglo-American physicist Freeman Dyson, who applied modern technology to it and suggested that an interstellar probe using photon pressure might be powered by either a visible-light laser or a maser (microwave laser). Then, in 1984, the physicist Robert L. Forward, using advanced computer technology, fielded the design of a gigantic light sail which he called Starwisp: a gossamer-thin aluminium film made from a hexagonal mesh of wire, 0.62 miles (1 km) in diameter and weighing only 0.7 oz (20 grams). There would be a micro-electronic circuit at each of the mesh's ten trillion intersections: semi-intelligent chips that would not only function as computer elements in a ten-trillion-component parallel-processing supercomputer, but would also be sensitive to light, each one acting as a tiny pinhole camera.

Forward's idea was that the ultra-delicate Starwisp would be built in deep space beyond the orbit of Mars. It would be powered by a 20-gigawatt (20 thousand million watts) microwave beam from a solar-powered satellite that would remain in Earth's orbit; the beam would be focused by means of a Fresnel zone lens, a lightweight structure consisting of concentric rings of ultra-thin plastic, of varying width, which alternate with empty rings. The radii of the rings would be adjusted so that the microwaves passing through the empty rings would be in phase at the focal point of the Fresnel zone lens. The latter would be huge, with a diameter of over 31,000 miles (50,000 km), about four times the diameter of Earth.

By means of photon pressure, the focused microwave beam would accelerate Starwisp to one-fifth the speed of light in a week. Seventeen years later, by which time the probe would have travelled three-quarters of the way to Proxima Centauri, mission control would send out a powerful pulse of energy which, travelling at the speed of light, would coincide with Starwisp's arrival at the Proxima Centauri system and switch on the craft's ten trillion microcircuits.

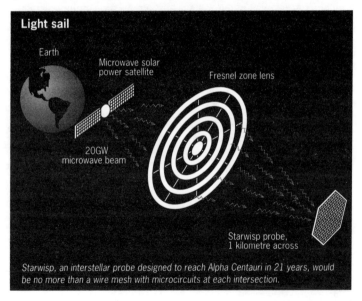

Light sail

Earth

Microwave solar power satellite

Fresnel zone lens

20GW microwave beam

Starwisp probe, 1 kilometre across

Starwisp, an interstellar probe designed to reach Alpha Centauri in 21 years, would be no more than a wire mesh with microcircuits at each intersection.

As Starwisp sped through the system at an incredible velocity of 37,500 miles/sec (60,000 km/s), its energized superchips would analyse the light coming from objects in the Proxima Centauri system and produce 25 high-resolution images per second; the mesh would then configure itself as a directional antenna and beam the data back to Earth. The probe would continue transmitting for four years, by which time it would be a light year past the Proxima Centauri system – a quarter of a century after it was launched.

If the probe detected planets that merited further investigation, the next step would be to send a larger probe equipped with better instrumentation. This would have a sail 2.2 miles (3.6 km) in diameter, made of aluminium foil about 16 nanometres thick. It would reflect 82 per cent of the light hitting it, allow 4.5 per cent of the light to pass through, and absorb the remaining 13.5 per cent. Together with its instrument package, it would have a mass of around 2200 lb (1000 kg). It would be powered by the photon pressure of a 65-gigawatt laser, which could either be in Earth's orbit or in orbit around the Sun, deriving energy from the powerful solar flux. The laser beam would be focused by a Fresnel zone lens 625 miles (1000 km) wide, stationed between the orbits of Saturn and Uranus.

Because of its greater mass, the larger probe would take twice as long as Starwisp to reach Proxima Centauri – about 40 years.

What is a quasar?

Quasars – the name is derived from Quasi-Stellar Object (QSO) – are the most distant extra-galactic objects known. They appear starlike, but they are not stars; they shine with the brightness of 100 galaxies, yet they are a million times smaller than our Milky Way.

The first of them was identified as a strong radio source in the constellation Virgo in 1962, and in August of that year Australian radio astronomers were able to fix its position very accurately. The source of the radio emissions seemed to coincide with a faint bluish star of magnitude 12.8, but when this object's spectrum was analysed, it became apparent that it was not a star. The spectrum revealed hydrogen lines with a marked red shift – a lengthening of the wavelengths of light that occurs when an object's motion is away from us, attributable to the Doppler effect – showing not only that this particular object was receding from our part of the universe at high velocity, but also that it was a tremendous distance away and extremely luminous.

Some quasars have been found to have a recessional velocity more than 90 per cent of the speed of light and are ten billion light years distant. One of them, discovered in April 1991 and designated BR 1202-07, is so far away that its light has been travelling towards us for most of the history of the universe.

Several theories have been put forward to explain the nature of quasars; the principal one is that they are supermassive black holes at the core of newborn galaxies, formed when a massive star explodes and collapses. As they suck in matter from surrounding space, they convert its gravitational energy into intense radiation. According to the astronomers who located it, the black hole in BR 1202-07 has a mass ten billion times that of the Sun and is swallowing matter at the rate of 100 solar masses a year.

Another theory suggests that a quasar could be a pair of huge gas clouds filled with charged particles and surrounded by a strong magnetic field. The clouds would trap electrons shooting out from a plasma nucleus. This in turn would result in the emission of radio waves, which might vary, or pulse as the nuclear 'generator' turned on and off.

Most astronomers support the theory that a quasar is the extremely luminous nucleus of a galaxy, whose stars are obscured by the quasar's brightness. The nucleus surrounds a black hole whose gravitational force, when exerted on nearby stars, tears them apart and sucks in their gas and matter. This flow produces energy that is seen as a quasar. No-one knows whether all galaxies went through a quasar phase, or just a fraction of them. If every galaxy did, then the Milky Way may contain a slumbering quasar at its heart – an object which 'switched off' when its fuel ran out.

Because of their extreme distance, quasars can act as useful tools in astronomical research. On their way to Earth, the light of the more distant quasars passes through hundreds of cold primordial clouds of hydrogen gas, relics of the Big Bang, and each one leaves its mark on the quasar's spectrum by absorbing light at a different frequency. Observations of this kind afford a glimpse of the primordial universe, when matter was more densely compacted.

What is gravity?

The story of Isaac Newton's apple is familiar to just about everyone. Newton is said to have sat in his garden in Kensington, London, after dinner one evening, and upon observing the fall of an apple, raised his eyes to the Moon and asked himself why that body did not fall to earth under the power of gravity, as the apple had done. Then he realised that the Moon was attracted to the Earth, and that only its orbital speed prevented it from falling, the centrifugal force of revolution counterbalancing the Earth's gravity.

Newton's apple tree was chopped down in 1820, 94 years after the event, but in some ways the force we call gravity remains as perplexing as it was to the great scientist-philosopher. Today, we know that it is one of the four fundamental forces, the others being electromagnetic, weak nuclear and strong nuclear, and although it is the weakest of the four, it acts over great distances. Just as the particle that is postulated as carrying the electromagnetic force is the photon, the particle associated with gravity is the graviton.

The Law of Universal Gravitation formulated by Newton more than ten years after the fall of the apple is stated as follows: Two physical points A and B of mass m and m^1 separated by a distance r exert an attraction on each other along the line AB; the attraction is proportional to the masses of these points and inversely proportional to the square of the distance between them. In accordance with this law, all objects fall to earth with the same acceleration, regardless of mass.

Gravity, then, may be defined simply as a force of attraction that arises between objects by virtue of their masses; on Earth, it is the force of attraction between any object in the Earth's gravitational field and the Earth itself.

There the matter rested until 1915, when Albert Einstein formulated his theory of gravity, more widely known as the general theory of relativity. Gravity, Einstein theorized, was the result of matter curving the fabric of space-time. Far from destroying Newton's law, Einstein's theory modified it and enabled it to be built upon, and when he drew electromagnetic radiation into the field of quantum mechanics, it might have been confidently expected that gravity would fit neatly into the unified field theory, which attempts to explain the four fundamental forces in terms of a single unified force.

However, it did not, and for years particle physicists have been trying to discover why. One of the principal viewpoints is that the attractive and far-reaching force of gravity is due to the exchange of the massless particles called gravitons; this is inspired by successes quantum theorists have had in viewing the fundamental forces as being the result of the exchange of 'carrier' particles – photons, for example, in the case of electromagnetism. All it needs, then, is to work out what gravitons do as they move about in space-time, and how they react with other particles, and you have them working as part of a unified field.

Then, in the 1980s, another mathematical theory came along to explain the behaviour of elementary particles like the electron, and physicists now believe that this could open the door to a valid quantum theory of gravity. Known as the string theory, it supposes that the fundamental objects in the universe are not point-like particles, but tiny stringlike objects. These objects exist in a universe of ten dimensions, although, for reasons not yet understood, only three space dimensions and one of time are discernible. The strings are under a colossal tension which keeps them tightly wrapped up.

To their surprise, particle physicists discovered that in mathematical equations, involving the string theory, gravity was an indispensable force and the graviton an indispensable particle. In 1984, the two scientists who had propounded this theory, Michael Green and John Schwarz, went a step further and combined string theory with a concept known as supersymmetry, in which they showed that supersymmetric strings – or superstrings – led to a theory which demands the existence of gravity without encountering the insoluble mathematical problems that had bedevilled every previous theory of quantum gravity. Within a year, scientists at Princeton University had discovered that superstring theory could unite gravity with the other three fundamental forces.

It may be only the beginning, yet already some physicists think that the superstring theory may be the ultimate 'theory of everything', explaining all the workings of the universe within one framework. It's a far cry from Isaac Newton's apple.

What are cosmic rays?

Since you started to read this sentence, about 50 cosmic rays have passed through your head. But don't worry: it happens all the time.

In 1912, an Austrian physicist named Victor Hess was measuring radiation levels with the aid of crude instruments called electrometers from the vantage point of a balloon when he discovered something very odd. He found that the level of radiation, after first declining, began to increase again as the balloon ascended. After some thought, he concluded that scientists needed to 'have recourse to a new hypothesis; either invoking the assumption of the presence at great altitudes of previously unknown matter, or the assumption of an extraterrestrial source of penetrating radiation'. In a time when scientists were easily held up to ridicule, it was a bold statement to make, and for years eminent scientists dismissed the idea of radiation penetrating the Earth's atmosphere from the cosmos as nonsense. It was nearly a quarter of a century before Hess was awarded the Nobel prize for his discovery.

Hess called this penetrating matter cosmic radiation, and although later work showed that most of the 'cosmic rays' penetrating the atmosphere were in fact particles – atomic nuclei and electrons – the name stuck.

Today, we know that cosmic radiation comprises streams of high-energy particles from outer space, consisting of protons, alpha particles and light nuclei which collide with atomic nuclei in the Earth's atmosphere to produce secondary nuclear particles that bombard the Earth in a continuous shower. These are mainly unstable subatomic particles called mesons, which are made up of two elementary and indivisible particles called quarks.

Protons – positively-charged subatomic particles which are

constituents of the nuclei of all atoms – make up 85 per cent of the cosmic radiation that reaches the Earth's atmosphere, the alpha particles and light nuclei accounting for the rest. Their density varies from 30 to 75 particles per cubic inch per second, and because the primary particles are electrically charged, their intensity at a given place depends on the strength of the Earth's magnetic field there.

Finding the source of cosmic rays has presented astrophysicists with not a few problems. Astronomers investigating light or radio waves, for example, are dealing with electromagnetic radiation that travels in a straight line, so their source is relatively easy to locate; cosmic rays do not. Because they have an electrical charge, the particles are affected by the tangled magnetic field in the vast gulfs between the stars, and as a consequence they pursue such tortuous paths that they reach the Earth with almost the same intensity from every direction. Also, low-energy cosmic particles are affected by the steady stream of particles and magnetic field from the Sun, the solar wind.

The key appears to lie in the study of gamma rays, quanta of electromagnetic radiation similar in nature to X-rays but of shorter wavelength, which are produced by cosmic ray particles when they impinge on the atoms in the hydrogen gas between the stars. Gamma rays are not affected by magnetic fields in space, and so reveal their origin directly. To put it simply, an excess of gamma rays at a particular point in the galaxy is good evidence that cosmic rays are coming from that direction.

Taking all available evidence into account, it seems that most of the lower-energy cosmic rays originate in the remnants of supernovae. Accelerated by the shock waves, they remain trapped within the remnant for a few hundred thousand years and then 'leak out' to diffuse across the galaxy. Higher-energy particles probably originate in neutron stars such as one component of the binary system Cygnus X-3, which is 35,000 light years distant and a major source of X-rays.

Only a few objects like Cygnus X-3 in the galaxy would be sufficient to produce all the high-energy cosmic rays.

Particles with even higher energies appear to come roughly from the direction of the Virgo cluster of galaxies, which contains the active galaxy M87. This has led some astronomers to speculate that the most energetic cosmic rays come from the vicinity of the active core of M87, which may contain a massive black hole.

What does a baby see?

The chances are that about half the readers of this book are wearing glasses or contact lenses. Although only about two per cent of human beings are born with vision defects, many of us need corrective lenses by middle age. Everyone becomes affected by presbyopia, which is the progressive loss of the power of accommodation for near vision and is due to age-related loss of elasticity of the eyes' crystalline lenses. It is usually noticed around the age of 45, when the eyes can no longer cope with normal reading distance.

Seeing is not simply a mechanical process, but a delicate balance between the eyes, emotions and brain. What we call sight results when visual impulses are transmitted via the optic nerves to the cortex, the part of the brain where they are processed, analysed and interpreted. Because the eyes are set apart, each one transmits a slightly different viewpoint, so the cortex combines the two into a three-dimensional image and also provides depth perception.

The process is automatic but somewhat complex, because each

eye is controlled by six muscles and all of them are brought into action when the eyes shift their focus. If we look at a close-up object the eyes converge, but if the object is distant the lines of vision of the eyes are parallel.

A human baby's eyes do not converge until it is about three months old, when it acquires this skill so that it can reach out to grasp objects. Before that, any grasping movement a baby might make – such as closing its hand around a parent's finger, for example – is purely instinctive and is a reaction to touch.

Up to three months, a baby will be visually aware of light and vague, unfocused shapes, but nothing more. The act of focusing its eyes, with all that that entails for its future life, is the most important thing that will have happened to it since it first drew breath, and psychiatrists now think that the first thing a child sees at the moment of eye convergence – which is usually its mother's face, and the expression on it – may have a profound effect on it subconsciously. For example, if the first impression is a disturbing one, the child may feel uncomfortable when examining close-up objects later in life, which may have an adverse effect on activities such as reading.

It is thought that about 75 per cent of children who fail to do well academically may have some form of undetected vision problem. The autonomic nervous system controls the body's glandular and digestive system as well as the focusing of the eyes, so if there is something wrong with the latter's function the whole body may feel disorientated and 'out of sorts'.

Statistics show that about five per cent of children under the age of ten are nearsighted, but at the age of 16 this rises to 20 per cent. The percentage of myopia (nearsightedness) among university students may be as high as 40, although many may be completely unaware that their vision is failing. In these cases, nearsightedness may be induced by concentrated close-up work over long periods. The key to healthy eyes is relaxation, with constant shifts of focus, and students – as well as others in occupations that require continual close-range work – who take regular physical exercise which involves looking at distant objects are far less at risk than those who do not.

The next time you look at a baby of six months or so, observe its eye movements. In one instant it will be gazing at something close – a toy perhaps – and in the next it will be staring into the far distance, completely ignoring its immediate surroundings. It is automatically flexing its eyes, a process that deserts us as we approach adulthood, mostly because we deliberately stifle it as our eyes strive to cope with the mass of information they need to assimilate as we pass through everyday life.

What is colour?

When white light, for example sunlight, falls on an object, some of the wavelengths of the light are absorbed and some reflected to the eye of an observer. The object appears to be coloured because of the mixture of wavelengths in the reflected light. For instance, a red object absorbs all wavelengths falling on it except those at the red end of the spectrum. Similarly, if you mix blue and yellow paints together, the result comes out green because between them the blue and yellow pigments absorb all the wavelengths except those associated with green.

When it comes to colour, our eyes have a tendency to mislead us simply because the human visual system is very sophisticated. True colour can be measured objectively by a physicist's spectrometer; it depends only on the relative amounts of light that an object reflects at all the wavelengths of the visible spectrum, which in turn is a function of the kinds and arrangements of the object's component molecules. An object that reflects a lot of 700-nanometer-wavelength light and little else is objectively red, while one that reflects only 520-nanometer-wavelength light is green.

Thanks to the brain's colour-processing apparatus, however, the colours we experience are different from those measured by physicists, because the brain takes into account its memory of what an object looks like (for example, an orange is orange, and not green, as we all know) and also scans the total amount of reflected light.

A good example of this is provided by a piece of chocolate and a piece of orange peel. We would obviously consider them to be of different colours, and yet an objective analysis shows that both reflect the same mix of light wavelengths. The only difference is that the chocolate, because of its molecular structure, reflects a smaller amount of light under a given level of illumination. If you compare a brightly lit piece of chocolate and a dimly lit orange, both will seem to be the same colour.

The first person to come to grips with the human eye's perception of colour was the English doctor Thomas Young (who, incidentally, deciphered the Rosetta Stone.) In 1807, he advanced the theory that the human eye had three kinds of receptor which gathered the whole range of reflected light. This theory was not fully confirmed until 1964, when American scientists succeeded in bouncing light off the cone-shaped light-detecting cells at the centre of the retina. They found that the cells were receptive to three distinct wavelengths corresponding to red, green and blue light.

A handful of people – perhaps one in 100 million – have only one functioning set of cone cells, so they are doomed to see everything in black and white. So do we all at night, because our cone cells do not function in very dim light. What happens when our eyes adapt themselves to the dark is that rod-shaped cells are activated around the periphery of the retina, and these provide us with our night vision.

So what happens when things appear to display colour at night? Say we look at Mars, the so-called 'Red Planet'. A spectral analysis would show that it reflects light of the same wavelength as the piece of chocolate, but seen against the backdrop of space it looks relatively brighter – so our brain deceives us into believing that it is orange, or red. The same effect tells us that the Moon is silvery-grey, when in fact it is dark grey.

For normal colour vision, all three sets of cone cells in the eye have to be working properly. When one of them is not, what usually happens is that it responds to a colour which is similar to the one that triggers its neighbouring set of cone cells, with the result that the brain becomes confused and is unable to interpret the signals properly. This is what we know as colour blindness.

What is an ulcer?

Anybody can get ulcers. In fact, one in six people in the western world suffers from them at some time. Fifty years ago, four males contracted them for every female, but that ratio is changing. Today, the number of women with ulcers is on the increase, and there are fewer male sufferers. No-one yet knows why, but one theory suggests that smoking may be a contributory factor.

Ulcers are caused when craters occur in the protective lining of the stomach (gastric ulcers) or small intestine (duodenal ulcers) as a result of an upset between the gastro-intestinal tract's natural defences and the substances that release the acid that is necessary in the digestion process. The substances involved are histamine, acetylcholine and gastrin, which are secreted in cells in the stomach lining. When the nervous system activates them, they join forces to release acid.

The problem is that they sometimes produce too much acid, which gradually begins to erode the walls of the stomach or duodenum. Hitherto, most treatments have depended on compounds designed to neutralize the acid – a bit like closing the stable door after the horse has bolted. But some acid is necessary to sterilize bacteria introduced into the body with food, so neutralizing it can have harmful side-effects. It is also needed for the activation of pepsin, the enzyne that breaks down proteins during digestion.

The real way to cure, scientists think, is to stimulate the gastro-intestinal tract's defence mechanisms. One of the most important of these is a group of fatty acids called prostaglandins, which are activated for a short time during the digestive process to inhibit the production of acid and also to stimulate the stomach's resistance. Tests on people with stomach ulcers (also known as peptic ulcers) have shown that their prostaglandin production level is low.

Another of the stomach's defences is a hormone called epidermal growth factor (EGF) which also assists in cutting down acid production. Researchers have been experimenting with synthetic products based on both hormones which can be taken in tablet or powder form. Other anti-ulcer drugs act directly on the substances that produce acid by 'taking over' their cells, and these can reduce the amount of acid generated by as much as 50 per cent. The surface lining of the stomach sheds 500,000 cells every minute and needs three days to renew itself completely, and since stomach acid can dissolve tissue in a matter of hours, such products give the stomach a fighting chance of repairing any damage.

However, the battle against ulcers is by no means won. Despite the efficiency of modern medicines, over half of all patients treated – excluding those who undergo surgery – get ulcers again within a year or two.

In their quest for a cure, scientists are having to revise their ideas about dieting. The membranes of the stomach lining's cells are composed of fatty substances known as lipids, and these can be penetrated by a wide range of agents which contain a low level of ionized molecules. They include vinegar, alcohol, some fruit juices and the detergents present in toothpaste. On the other hand, spicy foods – once thought to provoke ulcers – contain a high level of ionization in their molecular structure, and are relatively safe to eat.

It used to be thought that people working under high pressure, with associated stress and anxiety factors, were more prone to ulcers than those with an easier lifestyle, but new discoveries in medical science are turning that theory upside down. Recent surveys of men and women with high-stress jobs, such as air traffic controllers, have shown that they are no more likely to contract ulcers than anyone else.

What makes the world go round?

Consider the planet Earth for what it is: a gigantic spacecraft, 7923 miles (12,756 km) in diameter at the equator, completing an almost circular orbit around the Sun in 365 days, 5 hours, 48 minutes and 46 seconds at an average speed of 66,600 mph (107,000 km/s). It rotates around its axis once every 23 hours, 56 minutes, 4.1 seconds, the rotational speed at the equator being 1046 mph (1674 km/h). The Earth's average distance from the Sun is 92,860,000 miles (149,500,000 km.) The whole assembly – Sun, Earth and the other planets, asteroids, comets and so on that comprise the solar system – is itself orbiting the centre of our Galaxy and is moving towards a point in space presently occupied by the star Vega in the constellation Lyra – although Vega won't be there when the solar system arrives in a few million years' time, of course, because it is itself on the move.

Some pretty complex celestial mechanics are consequently involved in the Earth's passage through space. Under the attraction of the Sun and planets, its movements lose the geometrical regularity which they would possess if they simply obeyed those well-documented laws. Sometimes accelerating, sometimes slowing down, rotating and wobbling on its axis, the Earth behaves rather like a soap bubble floating through the air, its movement susceptible to all sorts of external influences.

As we said, the Sun is moving around the galaxy accompanied by its planets. This proper motion is influenced by two very unequal component forces, the first determined by the rotation of the galaxy and the second by the Sun's motion in relation to neighbouring stars. This second movement occurs obliquely to the plane of the solar system, and since the Sun itself is moving while the Earth goes round it, the Earth actually moves along a spiral – or more correctly, helical – path in space with reference to the surrounding stars. Seen from far out in space, the Earth's path would appear rather like a gigantic corkscrew about 186 million miles (299 million km) in diameter and with a 'pitch' of about 590 million miles (949 million km), the length of the planet's annual journey with the Sun.

As to the question of what makes the world go round, the rotation of the Earth – and of all other celestial bodies, for that matter – goes back to the time when the galaxy and the solar system were formed by the condensation of a rotating mass of gas. The preservation of angular momentum, the product of the moment of inertia and its angular velocity, would mean that any bodies formed from the gas would themselves be rotating. This rotation eventually slows down, but as friction and other influencing forces in space are very small, the process is very gradual. That it does happen, though, is apparent in the case of the Moon, whose rotational period has slowed down over aeons of time under the process of tidal friction, the gravitational interaction with its parent planet. It now spins on its axis once every 27.32 days, keeping the same face turned towards us as its rotational period coincides with the time it takes to complete its orbit around the Earth; the period is called a sidereal month.

What is arsenic?

The German alchemist Albertus Magnus (1193-1280) is credited with the discovery of arsenic. He also found that the steel-grey, brittle crystalline element could kill people very quickly and effectively, if administered in anything but tiny doses. Because of its lethal properties, Albertus Magnus and his fellow mediaeval alchemists kept arsenic a closely-guarded secret.

Arsenic is present in many ores, the soil, the sea and the human body. An average person's body contains about ten milligrams, and it is continually being extracted by the liver and converted into a form of acid which is expelled through the body's natural functions. The arsenic level is kept topped up through the food we eat. Everything we consume contains some arsenic, and some foods such as shellfish and the fish that feed on them can have relatively high levels, although these are still much lower than the amounts that would be needed to harm them or us. In fact, research has shown that when some animals such as chickens and goats are deprived of arsenic in their diet, they suffer from stunted growth until the element is restored. Scientists have also found that arsenic stimulates the production of haemoglobin, the respiratory pigment present in red blood cells that carries oxygen around the body. For this reason it was often prescribed as a treatment for anaemia before more effective medicines were found.

It was not until scientific research took on an organized aspect in the 18th century that the myths and mysteries surrounding the element were stripped away. At that time, doctors used arsenic compounds to treat rheumatism, arthritis, asthma, tuberculosis, diabetes and venereal diseases. An arsenic-based drug called Salvarsan, discovered by the German chemist Paul Ehrlich in 1909, remained the primary cure for syphilis until the advent of penicillin in the 1940s.

Perhaps the most famous exponent of arsenic as a medicine was Dr Thomas Fowler, of Stafford Infirmary, who was so impressed by the effectiveness of a medicine called Tasteless Ague and Fever drops that he and a fellow doctor named Hughes decided to analyse it. They found that the active ingredient was arsenic. Together, they devised their own recipe using arsenic trioxide and potassium carbonate – which they knew as white arsenic and potash – and marketed it as Fowler's Solution. It made Fowler famous; what Hughes got out of the deal is not recorded.

But it was the darker side of arsenic that made the element notorious. The first recorded attempt at murder using arsenic occurred in 1384, when conspirators made an unsuccessful bid to poison King Charles IV of France, and the infamous Borgias – Cesare and Lucrezia – are known to have killed many people with it. Another notorious arsenic poisoner was a woman called Toffana of Sicily, who was a principal supplier for half a century. Her arsenic-based poison, known as 'Manna of St Nicholas', claimed at least 500 victims, including two popes. Her deadly trade was uncovered in 1709 and she was put to death.

Arsenic poisoners of the 18th century were protected by the fact that the symptoms produced in the victim could easily be mistaken for those of various ailments. It was not difficult to administer, and only a small amount in a glass of wine was enough to do the job. By the middle of the 18th century, though, forensic science – although still in its infancy – was often able to show that arsenic had been responsible for a person's demise. In 1752, Mary Blandy – who had murdered her father with arsenic – became the first person to be convicted and executed on such evidence.

Cases of murder by arsenic were rife in the 19th century, but there were also many cases of accidental poisoning. To this day, there is controversy over whether Napoleon Bonaparte was deliberately poisoned by his English captors on St Helena, or whether he absorbed a lethal dose of arsenic from a wallpaper dye called Scheele's green, in common use at that time. Flypapers containing arsenic were also responsible for a number of deaths. In 1858, 200 people were taken seriously ill in Bradford and 20 of them died as a result of eating peppermint lozenges containing arsenic, mistaken for harmless calcium sulphate by the confectioner. And in 1901, 6000 people in Manchester were poisoned when they drank beer containing arsenic that had been produced in the brewing process. The death toll reached 70. After this disaster, a Royal commission was set up to investigate the amount of arsenic in beer, glycerine, glucose, malt and treacle, resulting in the imposition of strict controls that are still in force today.

What is a snore?

Most people indignantly deny that they snore. In fact everyone snores to some extent, because the upper airway at the back of the throat – a tube lined with various muscles, including those of the tongue and soft palate – becomes relaxed during sleep, so that the airway sags inwards every time the sleeper breathes in. If the mouth is open, this sets up a vibration and the result is the sound of snoring.

Heavy snoring can have serious consequences. In some cases, it can cause the upper airway to collapse completely, cutting off the flow of air. When this happens the sleeper automatically struggles to regain breath. The rib cage and diaphragm heave, bringing about abnormal changes in air pressure within the lungs. The flow of blood within the chest, heart and lungs is impaired. Blood pressure rises sharply, the heart begins to beat erratically, and levels of oxygen in the blood fall, affecting the heart's function even more.

This is the point at which snoring becomes dangerous, especially to people suffering from heart disease or blood circulation problems. Scientists have established that more people die in the seemingly tranquil period between five and six o'clock in the morning than at any other time, and in many cases death at that hour may be due to snoring and other breathing disorders that result in a dangerously low level of oxygen intake.

In the case of healthy people, this state of suffocation lasts for about 15 seconds, at which point the centres in the brain that control breathing alert the rest of the brain to the danger and the sleeper begins to wake up – usually with a loud explosion of breath known as the 'heroic snore'. The person may continue to snore at a lower level, but returns to normal sleep as breath is restored and oxygen levels in the blood rise. The recovery takes only about ten seconds, so that the whole cycle lasts less than half a minute from start to finish. However, the process repeats itself many times – as much as several hundred times in a single night, in very severe cases. Oddly enough, the snorer usually reports having slept well, without realizing that their sleep pattern has been severely disrupted, and is unable to explain why tiredness sets in during the day. According to most surveys, one in three people do not get a good night's sleep because of their own snoring.

Young people appear to suffer more from snoring than those over the age of 35. One survey at Edinburgh University showed that nearly twice as many young men and women complained of too little sleep than those over 35. One possible reason is that under-35s may consume more alcohol than those outside their age group. Alcohol has an adverse effect on the respiratory centres in the brain and reduces blood flow to the extent that some neurons do not get the oxygen they need to function properly. The body of the sleeping person tries to compensate for this by increasing the intake of breath, leading to the severe snoring condition that can cause the collapse of the upper airway. Excessive smoking has a similar effect.

You can't stop yourself snoring, as you are asleep when it happens, but you can take steps to prevent it. Apart from obvious measures such as cutting out smoking and reducing alcohol consumption in the evening, you can take some form of exercise before you go to bed. Even a brisk walk around the block will give your body's oxygen intake a big boost and help to induce healthy sleep, as long as you don't make yourself over-tired.

A curb on snoring will doubtless improve relations with your partner, too. The volume of some people's snoring can reach 70 decibels, which is more intense than than a road drill.

Snoring certainly annoyed a man called John Wesley Hardin, who was one of the Wild West's more notorious gunfighters and who took a rather extreme measure to cure the snoring of a man in the hotel room next to his. He went in and shot the man dead.

What are crop circles?

A crop circle is a circular area of flattened grain. The phenomenon is found mostly, but not exclusively, in southern England, and about 1000 are reported every year.

In fact, mysterious circles in corn and other vegetation have been around for a long time, but it was not until the second half of the 20th century that they received any media coverage. The reason was the big upsurge of interest in unidentified flying objects (UFOs), spurred by the publication in the 1950s of numerous popular 'flying saucer' books. All of a sudden, the strange circles in the corn were transformed into UFO landing sites, and even now no amount of reasoned argument can convince believers that they are anything other.

So what are crop circles, if not extraterrestrial calling cards? Regrettably, it has to be said that some of them are admitted hoaxes. When a large circle surrounded by four smaller and equally-spaced circles was found at Westbury, Wiltshire, in July 1983, a local surveyor assisted by a group of newsmen showed how easy it was to perpetrate a crop circle hoax with the aid of a long chain fixed at a central point and then rotated around the circumference. The two groups of circles were indistinguishable.

Most of the research into crop circles has been carried out by dedicated amateur investigators rather than by scientists, but physicists who have studied the phenomenon are convinced that there are several valid meteorological and other physical explanations for them. Natural events that might account for crop circles include fungi, plant and grass sickness, lightning, animal behaviour, whirlwinds, tornadoes, rain, helicopter rotor downwash, exfoliation, geomorphological features and so on.

Some scientists suggest that they are created by an electromagnetic whirlwind or 'plasma vortex', and Italian researcher Maurizio Verga, who has specialized in investigating physical traces of alleged UFO sightings, cites numerous cases where this and the tornado theories hold good. In one case, 11 circles ranging in diameter from 8 to 20 ft (2.5 to 6 m) were discovered in a wheatfield near Rossburn, Canada. Some were aligned directly under high tension cables. The flattened grain in the circles appeared to have been scorched, but analysis showed that the darkening was in fact due to mould, caused by the grain lying flat against wet ground. The phenomenon was almost certainly due to the effect of a tornado; this area was particularly prone to them, and a month or two earlier the weather conditions had been just right to produce them. Another series of circles, found in New South Wales, Australia, in 1970, were caused by primary and secondary lightning strikes during a coinciding period of electrical storms.

Some crop circles are undoubtedly caused by ball lightning, dealt with elsewhere in this book. Eyewitnesses linking 'UFOs' with crop circles have described seeing balls of light emitting a curious hissing sound, which is a characteristic of ball lightning as it follows a spiral path to earth.

What is the horizon?

The word horizon is derived from the Greek *orizo*, meaning a bound or limit. We apply the term to the line all around us where the land or sea meets the sky. Because of their apparent diurnal motion, the Sun, Moon and stars move in relation to the horizon; as they rise and set, they come into view or disappear round the limit of that part of the terrestrial globe which is visible to us.

In astronomical works the Earth is shown to be a sphere, and the eye of the observer is assumed to be at surface level of the globe. In fact this is never the case, as an observer is always at some height above the surface of the sea or land, and the 'offing', or distance to the horizon, depends on this factor. A sailor at a masthead, for example, sees a good deal farther than one on deck. The question is, how far?

If your eyes were exactly at sea level, your horizon would be directly in front of you, and you would see no distance at all. Theoretically, the curvature of the globe means that the surface would fall away from you, immediately beneath a horizontal plane tangent at that spot. This plane is the astronomical horizon, and is merely a boundary line; the half of the celestial sphere which happens to be above it is visible, while the other half is invisible.

But what if you were to stand up in a boat? Assuming that your eyes were now 6 ft (2 m) above sea level, the horizon would be almost exactly 3 miles (5 km) away, with the visible distance increasing the higher you ascend.

You can work out the distance from your eyes to the horizon by means of a simple formula. Take the distance in feet to eye level, multiply by three then divide by two. Now take the square root of your answer, and the result gives you the distance in miles to the horizon. For example, if you are standing on a clifftop and your eyes are 100 ft (30 m) above sea level, then $100 \times 3 : 2 = 150$, and the square root of 150 is 12.25, or near enough – so the horizon is 12.25 miles (19.7 km) away. Admittedly, this formula is accurate only at low altitudes; the higher you go, the more invalid it becomes. Although the horizon recedes very rapidly at first as the observer's altitude increases, the rate of recession thereafter becomes very much slower. If you ascend from sea level to 3300 ft (1000 m) the distance of the horizon increases from zero to 70.6 miles (113 km), but if you ascend from 29,500 ft to 32,800 ft (9000 to 10,000 m) the distance to the horizon would expand only from 210 to 223 miles (338 to 357 km). If you continue to travel away from the Earth, the change of angle of the line joining the eye to the horizon, which is a tangent, becomes progressively smaller, so that eventually you will reach a stage where you double the distance without perceptibly increasing the amount of Earth you can see.

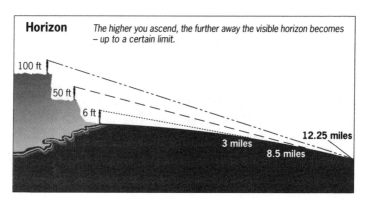

Horizon *The higher you ascend, the further away the visible horizon becomes – up to a certain limit.*

100 ft

50 ft

6 ft

3 miles

8.5 miles

12.25 miles

To find out the area you can scan, take the distance to the horizon and multiply it by itself – in other words, square it. Now multiply it by 3.14, and the answer gives you the area you can see in square miles. A person standing upright at sea level, with the horizon 3 miles (4.8 km) distant, will cover about 28.25 sq miles (73.17 km^2)

What is depression?

Most of us experience these feelings at some time in our lives: sadness, hopelessness, pessimism, a general loss of interest in life. It is what we call depression.

Luckily, for the vast majority, the experience is only an occasional one, and occurs as a natural response to a particular event – the death of a close relative for example. But depression remains a very common disorder, affecting one in ten men and one in five women at some point.

It is the elusive quality of depression that makes it so difficult to deal with, and the disorder so sinister. The pain, though quite real, is not caused by microorganisms; the suffering is palpable, but few potions can relieve it. Yet little by little, scientific research is stripping away the fog that surrounds the disorder, and psychiatrists are discovering that depression has its roots not so much in emotional turmoil as in recognizable biochemical imbalances.

It has been found, for instance, that some types of depression can be diagnosed simply by studying irregularities in a patient's blood platelets – tiny membrane-bound cell fragments that bud off from large cells in the bone marrow and help the blood to clot. Freckling the surface of platelets is a complex substance called serotonin, which constricts small blood vessels at the site of bleeding, so reducing blood loss; it is also present in the brain, where it acts as a neurotransmitter, regulating mood and other diverse functions.

Serotonin has often been called the 'molecule of depression', because some clinically depressed people have inadequate amounts of the substance in their brains. Research has shown that this insufficiency is accompanied by a decrease in the 'uptake sites' in the brain where serotonin attaches itself to the blood platelets. In some depressed patients, the total of serotonin uptake sites may be 30 to 40 per cent lower than average. Because reduced levels are present even when a person is not depressed, testing a patient's platelets can serve to identify not only those people already suffering from depression, but also those likely to develop it.

Some researchers believe that suicidal tendencies in depressive patients can be detected by testing the patient's spinal fluid for 5-HIAA, a metabolic by-product of serotonin. They base this theory on Sigmund Freud's interpretation of suicide as an act of violence directed inwards, and think that depressed patients who have made one or more suicide attempts might have too little 5-HIAA, indicating a serotonin deficiency. In one study carried out in Sweden, over 65 per

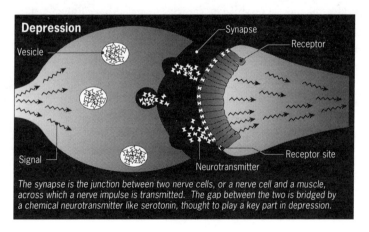

The synapse is the junction between two nerve cells, or a nerve cell and a muscle, across which a nerve impulse is transmitted. The gap between the two is bridged by a chemical neurotransmitter like serotonin, thought to play a key part in depression.

cent of people who had attempted suicide were shown to have below-average levels of 5-HIAA.

Research of this kind has led to the development of medication designed to offset serotonin deficiences. However, although such developments are encouraging, they do not provide a complete set of answers to the question of depression. Even as the arsenal of anti-depressant weaponry grows, different causes of the disorder are coming to light all the time, and each one requires a completely different treatment.

Scientific research in the 1980s discovered that the same type of human leukocyte antigen (HLA) – a blood protein produced by the immune system – seems to appear in clusters within families, indicating that it is genetically transmitted. This deviates from the generally random pattern of inheritance, and it has been shown that many family members who carry HLA develop depression.

Somehow the two are linked, and a possible explanation is that genes for depression and those dictating the HLA makeup are in close enough chromosomal proximity that they are inherited together. On the other hand, the HLA genes alone may pass from parent to child and in turn contribute to chemical imbalances that 'switch on' depression.

What is a hobby-horse?

The hobby-horse, which is still trotted out around the British Isles at various times of the year, is as old as British history. Its origins go back to the ancient Celtic festival of Beltane, which was held on May Day, considered by the Celts to be the first day of summer. The festival was accompanied by animal sacrifices (and probably human ones too), and in this ritual it was the horse, greatly revered as a totem among the Celtic tribes, which had pride of place.

The pagan connotations of the hobby-horse have long since gone, and the serious sacrificial ritual has been replaced by a more carnival-like atmosphere, but in some places there are still strong echoes of the pagan past. This is particularly true of the west country, where Celtic bloodlines and traditions still run strong.

Minehead, in Somerset, has two hobby-horses, both 8 ft (2 m) long ribbed canvas structures, rather like upturned boats, festooned with ribbons. One, the Town Horse, is more modern than the other, which is known as the Sailor's Horse; the latter used to have a horse's head with snapping jaws, but it now boasts a high, pointed cap and a fearsome mask, concealing the head of the man inside who provides the legpower. Local legend has it that the Sailor's Horse commemorates a sea monster that emerged to rout a band of marauding Vikings a thousand years ago, but its true origins lie in the Beltane ritual.

The hobby-horses prance around the area from 30 April to 3 May each year, gyrating to the rhythm of drums and melodeons. They also tour the neighbouring town of Dunster, bestowing good luck on the inhabitants for the coming year. Once, they were accompanied by attendants called Gullivers, who used to extort money from people;

anyone who refused to pay up was beaten on the buttocks with an old boot. This custom got a bit out of hand in the 19th century, and was discontinued after a man was enthusiastically beaten to death.

Padstow, in Cornwall, also has two hobby-horses, but here the ritual has more pagan undercurrents. The festival starts in the early hours of May Day with the singing of the May Song to rouse the inhabitants; then, at 10 am, the Blue Ribbon or Temperance 'Oss comes out, and is followed an hour later by the Red Ribbon 'Obby 'Oss. Each 'Oss consists of a circular frame covered by black material, with a masked head in the middle; the Red Ribbon 'Oss, which lives in the Golden Lion Inn, is much the older of the two, and attracts most of the limelight.

The 'Osses are accompanied by Teasers, who carry sticks to goad them into more gyrations as they prance through the town, mostly chasing girls. If an 'Oss catches a girl and conceals her under his skirt, she will have a child within the year if she is married, or be married within the year if she isn't – a clear throwback to ancient fertility rites. Until the 1930s there was also a ritual at a nearby pool, where the 'Oss took up some water and sprinkled his followers – a re-enactment, perhaps, of an ancient rain-making ceremony. Every so often the 'Oss dies and comes back to life again, which is clearly symbolic of the old Celtic year giving way to the new. The dancing goes on into the afternoon, when the two 'Osses meet at the maypole.

Another feature of the Padstow festival is that the people carry sprigs of greenery, which was also a feature of the Celtic ritual.

In some parts of Britain, the hobby-horse has been transposed to the festival of Christmas, with mummers wearing horse costumes and acting out the life and death ritual, and to Whitsuntide. Whatever the time of year, the ancient Celtic totem is very much alive and kicking.

What is a will-o'-the-wisp?

For centuries, travellers picking their way across marshy moorland at night were puzzled and frightened by strange balls of cold light that seemed to dance over the surface. It was said that the lights were evil spirits, sent to lure the unwary from the beaten track into the clinging morass.

Today, science has a much more rational explanation. The lights are caused by methane, the colourless, odourless gas that is lighter than air and explodes into a bluish flame when it comes into contact with oxygen. It is the chief constituent of natural gas, and is also present in firedamp, the gas that occurs in coal mines. It is produced by rotting vegetation, which ignites by spontaneous combustion to become the pale flame popularly known as a will - o'-the-wisp.

The study of methane and how it behaves has become very important in recent years, because it accounts for about 38 per cent of the warming of the Earth through the greenhouse effect. Some 15 per cent of the methane in the atmosphere is produced by the digestive gases of cattle, and according to some estimates the total amount is likely to double over the next half-century.

What is photosynthesis?

Life on Earth as we know it could not exist without carbohydrates, the chemical compounds composed of carbon, hydrogen and oxygen which provide energy for all living things. They are, if you like, life's working capital, and they in turn would not exist without photosynthesis.

Every year, through photosynthesis, the green plants of Earth convert solar energy into a staggering 200 billion tonnes of carbohydrates. In the process, they produce another commodity that is equally essential to most living organisms: oxygen. Photosynthesis, therefore, is the ultimate source of the food we eat and the air we breathe.

Every biology student knows how the process works. A photon of solar energy excites a molecule of chlorophyll, the green pigment present in plant cells called chloroplasts, and causes it to give off an electron. The chlorophyll molecule, now bearing a positive charge, picks up a replacement electron from water; the water is split into its component atoms in the process, producing oxygen and positively charged hydrogen ions. The positive ions and negative electrons are transported to opposite sides of a chloroplast membrane, creating an electrochemical potential similar to that inside a battery. The energy so created then synthesizes molecules of adenosine triphosphate (ATP). This is a nucleotide molecule which is found in all cells and yields large amounts of energy, motivating the thousands of biological processes needed to sustain life, growth, movement and reproduction. In this case, it converts carbon dioxide into carbohydrate.

This is the basic explanation of photosynthesis, but far more intricate processes lie behind it all, and scientists are still investigating a number of fundamental questions. For example, how are electrons and hydrogen ions passed from molecule to molecule within chloroplast membranes? What is the structure of the reaction centres in which key elements of photosynthesis take place? How is water split during the photosynthesis process, and how are all the necessary processes controlled by a plant's genes?

Discoveries in recent years are slowly unlocking some of the answers. For example, a key molecule in electron transmission in plants, called quinone, appears to perform an energy transfer function in all living things, and all energy-coupling systems – the mechanisms that transport electrons and use some of their energy to make ATP – use quinones.

Photosynthesis in the Carbon cycle
CO2 in atmosphere
Plant respiration
Vegetation burning   **Photosynthesis**
Grazing
Death
Microbial respiration
Fossil fuel combustion
Microbial biomass
Stabilised soil carbon
Fossil fuels

The continuation of life on Earth depends on the carbon cycle, in which photosynthesis plays a key role.

Scientists have not yet been able to reproduce the photosynthetic process artificially, but they have gone a long way towards it. In one experiment, a chlorophyll-like molecule called porphyrin is linked to a quinone molecule; when the porphyrin is struck by light, it donates an electron to the quinone. The electron's natural tendency is to jump back immediately, giving up its energy as heat. In a similar manner, electrons produced during photosynthesis want to jump back to chlorophyll before they can participate in ATP synthesis, but they are prevented from doing so by a structure of proteins. The whole process happens very quickly; the initial charge separation occurs in less than a billionth of a second, a speed that compares well with the function of modern computers.

What is skin?

If anyone were to ask you what is the largest organ of your body, you might have to stop and think for a bit, and even then you would probably come up with the wrong answer. It is in fact your skin, which – although only about three millimetres thick – covers an area of 21 sq ft (2 m²) on average.

A masterpiece of engineering, the skin performs two main functions: keeping your body's water in, and harmful chemicals out. When the remote ancestors of today's mammals first crawled out of the primaeval oceans of Earth on to dry land, they were protected by a thick, scaly outer covering, guarding them against the hostile environment; mammalian skin, with its coats or tufts of hair, evolved about 150 million years ago.

The skin is laminated: that is, it is built up in layers. There are two main ones, the outermost of which is called the epidermis. It consists of cells known as keratinocytes, which produce a family of fibrous proteins called keratin. The name is derived from the Greek *keratos*, which means horn. If pressure is put on some parts of the skin, more keratin is produced, which is why we have thicker skin layers on the soles of our feet and the palms of our hands. Our fingernails, toenails

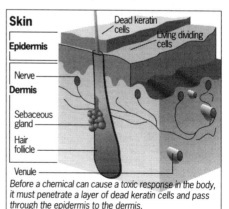

Skin

Dead keratin cells

Living dividing cells

Epidermis

Nerve

Dermis

Sebaceous gland

Hair follicle

Venule

Before a chemical can cause a toxic response in the body, it must penetrate a layer of dead keratin cells and pass through the epidermis to the dermis.

and hair are particularly rich in keratin, as are the hoofs, horns, claws and feathers of other creatures.

When keratinocytes die off, they form an outer layer of skin called the *stratum corneum*, which is waterproof and which prevents us from losing too much of our body's vital water. It also protects us from chemicals and other potentially damaging external agents, and in the process it is constantly being sloughed off. The lost cells are replaced by more keratinocytes which are produced deep down in the epidermis and gradually move up to the surface. We shed our entire skin surface every month or so, which accounts for some of the dust that builds up in our homes.

In its protective role, the skin contains many important enzymes (biological catalysts produced in cells which are capable of speeding up the chemical reactions necessary to sustain life by converting one molecule into another) that can break down or inactivate toxic chemicals. On the other hand, certain enzymes may activate chemicals or make them more toxic. As long ago as 1775, an English doctor named Percivall Pott noticed an increase in scrotal cancer among chimney sweeps; this was due, as is now known, to skin contact with the polycyclic aromatic hydrocarbons in soot. The hydrocarbons themselves are harmless, but the cytochrome P450 enzymes in the skin convert them into reactive compounds that can damage cellular DNA and cause cancer.

The strength and elasticity of the skin is due to the dermis, the bottom layer that lies beneath the epidermis. A thicker layer of skin cells, it comprises a tightly interwoven mesh of strong and elastic protein fibres. Most of these belong to a family of proteins called collagens, interlaced with another group known as elastin. The fibres are embedded in sugar-like molecules named glycosaminoglycans. It is collagens which make leather so durable.

Woven throughout the tissue of the skin are the blood vessels, lymph vessels and nerve endings, as well as cells called fibroblasts which produce collagen and elastin. The blood supply provides nutrition and energy to the dermis and epidermis, while the specialized nerve endings enable us to feel heat, cold, pain and provide the sensation of touch. Other cells provide an early warning network to the immune system, alerting it to the presence of agents that might cause inflammatory skin diseases such as eczema (also known as dermatitis) which inflames the epidermis and upper dermis and causes severe itching. Dandruff is a mild form of this condition, which is caused by an allergic reaction to chemicals.

One little mystery about the skin concerns a greasy substance called sebum, which is produced by sebaceous glands attached to hair follicles. Nobody knows why it is there, but it blocks the sebaceous gland's duct, leading to the skin surface, and gives rise to that teenage misery, acne. A partial blockage results in blackheads, which have a dark plug of keratin and sebum at their centre.

All over the skin of humans and other primates, sweat glands produce surface perspiration designed to keep the body at an acceptable temperature. In most other animals they occur only around the feet and face areas, which is why dogs pant and pigs wallow in wet patches. In humans, sweat glands occur in greater numbers in the male than in the female; their secondary function is thought to be to communicate social and sexual messages. What they often communicate is an unpleasant odour, caused by organic matter secreted in the apocrine glands and broken down by bacteria.

What are vitamins?

For hundreds of years, sailors had a major problem. On long voyages they tended to suffer from a condition that started with general weakness, progressing through aching of the joints and muscles, followed by bleeding of the gums, and the drying up of skin and hair. All too often, men died of it. In the 18th century the Royal Navy began to provision its ships with barrels of lime juice, with which its seamen were dosed on a daily basis, providing them with both an antidote and a derogatory nickname bestowed on them by the Americans.

Many years later, it was discovered that the scourge of the high seas, scurvy, was caused by a deficiency of vitamin C. This is ascorbic acid, which is present in fresh fruit and vegetables and is one of several chemically unrelated organic compounds. They are present in the body in small quantities and are necessary for its normal, healthy functioning.

The existence of such compounds was proposed by a Polish-born biochemist, Casimir Funk, in 1912, but it was another three years before the fact that certain deficiency diseases could be remedied by providing the patient with extracts from other foods became firmly established. By this time it was known that two groups of 'vitamines' (the word was coined by Funk from 'Vitalamines') were involved, one being water-soluble and present in such things as yeast, rice polishings and wheat germ, the other being fat-soluble and present in egg yolk, butter and fish-liver oils.

The water-soluble substance, which was already known to be effective against beriberi – a tropical metabolic disorder accompanied by an inflammation of the nerve endings and caused by a lack of thiamine, the vitamin found in seeds and grain – was given the name vitamin B, while the fat-soluble group was called vitamin A. The vitamin B complex was later subdivided as more was learned about its properties, and other groups were added to the list.

Vitamin A is essential for normal growth, for the formation of bones and teeth, for cell structure, night vision, and for protecting the linings of the respiratory, digestive and urinary tracts against infection. It is absorbed by the body in the form of retinol, found in the substances mentioned above.

The second vitamin complex comprises thiamine (vitamin B1), already mentioned; riboflavin (vitamin B2), niacin, pantothenic acid and pyridoxine (vitamin B6), cyanocobalamin (vitamin B12), biotin (vitamin H) and folic acid. One of the most important of these is vitamin B12, which plays a vital role in the activity of certain enzymes and in the production of the genetic material of cells, especially the red cells of the bone marrow, of carbohydrates, and in the functioning of the nervous system. The main effects of vitamin B12 deficiency are anaemia, a soreness of the mouth and tongue, and numbness and tingling of the limbs caused by damage to the spinal cord. There may also be depression and loss of memory.

Not only humans need vitamins; all animals do too, although not necessarily the same ones. For example, choline is a vitamin essential to rats and some birds, but they are unable to produce sufficient amounts for themselves. It is present in some fats and in egg yolk, and is the reason why bluetits peck the tops off milk bottles in order to get at the cream.

What is a virus?

In the latter half of the 18th century, a British doctor called Edward Jenner overheard a milkmaid claim that she could not contract smallpox – at that time a scourge of mankind – because she had already contracted cowpox, a bovine disease that caused only mild symptoms in humans. Jenner decided to put the girl's claim to the test, and in 1796 he carried out his famous experiment in which he inoculated eight-year-old James Phipps with liquid from a pustule on the hand of Sarah Nelmes, who had cowpox. The experiment was a success, but it was nearly half a century before the English Parliament passed a law making vaccination (a word derived from the Latin *vaccina*, cowpox) compulsory.

Jenner had no means of knowing it, but he had hit upon the human immune system's ability to respond to and counter microorganisms which it had encountered before. Such microorganisms include viruses, bacteria, fungi and protozoa.

Of the four, viruses are the smallest and most difficult to detect. They are anything up to one-hundredth the size of the smallest bacteria, from which they differ in that they have a much simpler structure and method of multiplication. Among the diseases they cause are canine distemper, chickenpox, the common cold, herpes, influenza, rabies, polio, yellow fever, AIDS, and a variety of plant diseases.

A virus is a particle consisting of a core of nucleic acid – either DNA or RNA – enclosed in one or two protective shells called capsids, which are made of protein. These capsids are built from a number of identical protein subunits arranged in a highly symmetrical form, either as a spiral tube or a 20-sided solid (icosahedron). Some viruses also have another layer of protein called the viral envelope, which surrounds the outer capsid to provide extra protection.

Viruses gain access to the body by every possible entry route: they are swallowed in food and fluids, inhaled in droplets, passed through punctured skin on the saliva of insects or animals, or on infected needles; via the mucous membranes of the genitalia during sexual intercourse; and by the conjunctiva of the eye after accidental contamination.

Viruses are only able to function and reproduce themselves if they can invade a living cell and use the cell's system to replicate. In the process they may disrupt or alter the host cell's DNA, possibly causing serious disease if vital organs are affected. Many viruses attack cells near the point of entry to the body; some enter the lymphatic vessels and spread to the lymph nodes, where many are overwhelmed by white blood cells called cytoxic T lymphocytes, but others pass to the blood and spread throughout the body in a matter of minutes.

A healthy human body reacts to viral invasion of cells by producing interferon, a cellular protein that makes up part of the body's defences against viral disease. Three types (alpha, beta and gamma) are produced by infected cells and enter the bloodstream and uninfected cells, making them immune to virus attack. This process may itself produce symptoms such as fever and fatigue.

The immune system deals fairly rapidly with most viruses. Sometimes, however, the viral attack is so swift that the immune system has no time to react adequately, and the result is often serious damage or even death. Sometimes, also, the virus evades the immune system or 'hides' from it, so that the infection becomes chronic or recurrent. This happens with the HIV virus, which invades and disrupts the activity of T-lymphocytes, so that these fail to produce defensive antibodies which normally come into play against a wide range of infections.

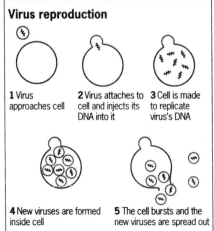

Virus reproduction

1 Virus approaches cell

2 Virus attaches to cell and injects its DNA into it

3 Cell is made to replicate virus's DNA

4 New viruses are formed inside cell

5 The cell bursts and the new viruses are spread out

No-one really knows where viruses originated. The general belief is that they are degenerate forms of life derived from cellular organisms, or pieces of nucleic acid that have broken away from the genome (full complement of genes) of some higher organism and taken up a parasitic existence.

What was the Black Death?

It was the worst disaster to befall Europe until the terrible wars of the 20th century. It began its relentless progress from the Gobi in central Asia, making its way along the trade routes from the Orient to the Near East and on through the great ports of Renaissance Italy and France into the heart of the continent.

In four years, from 1347 to 1351, it killed as many as 28 million people. In the cities, where vermin were rampant and sanitation primitive, mortality rose as high as 60 per cent. In London, with its population of no more than 50,000, people were dying at an average rate of 290 a day in the summer of 1349. The bodies of the dead lay unburied; no-one, not even the priests, would venture near them. Terror degenerated into hysteria and anarchy; religious minorities were persecuted.

Then, as suddenly as it had arrived, the terrible plague known as the Black Death departed, leaving the population of Europe decimated and its social and economic face changed for ever.

Before the middle of the 6th century AD, smallpox and measles were the biggest killers in the history of plague. Measles was then extremely virulent, and in one epidemic it killed 5000 people a day in

Rome. Then, in 541, the earlier plagues were overshadowed by the arrival of a third disease, caused by a complex series of bacterial strains that would be known many centuries later as *Yersinia pestis*. It lives in the digestive tract of fleas, particularly rat fleas, but if these hosts are not available it can exist in the human flea, *Pulex irritans*. Periodically, and for reasons epidemiologists still do not fully understand, the bacilli multiply in the flea's stomach in sufficient numbers to cause a blockage, so threatening the flea with starvation. While attempting to feed, the flea then regurgitates large numbers of *Y. pestis* bacilli into its victim through a puncture in the skin. This process is crucial to the plague's progress, as *Y. pestis* cannot pass through healthy skin.

Dozens of rodent species carry the plague, but in Europe the most important carrier was the black rat, which lived in or close to human habitation. However, the fleas responsible were also carried by other hosts, including virtually all household and barnyard animals except the horse, whose odour the fleas apparently found repellent. It was only after the supply of animal hosts was greatly depleted that the fleas carrying *Y. pestis* turned to humans.

The most common manifestation of the plague was in its bubonic form, which is still rife in some areas of the Far East. It takes about six days to incubate from the time of infection, and the first symptom is generally a blackish, often gangrenous pustule at the point of the bite. This is followed by an enlargement of the lymph nodes in the armpit, groin or neck, depending on the place of the bite, forming the 'buboes' from which the disease takes its name. The next stage is subcutaneous bleeding, causing purplish blotches. This bleeding causes cell necrosis and intoxication of the nervous system, leading to psychological and neurological disorders. Although bubonic plague is the least lethal of all plague types, it still kills between 50 and 60 per cent of its victims.

The second form, pneumonic plague, is far more lethal and can be spread directly from victim to victim. It infects the lungs, and after an incubation period of two or three days there is a rapid fall in body temperature followed by severe coughing and a discharge of bloody sputum that contains *Y. pestis* bacilli, so that transmission is airborne. Infection is followed by a paralysis of the nervous system and coma, and the fatality rate is 95 to 100 per cent. It was this form of plague that ravaged England, and particularly London (1964-65); it gave rise to the rather macabre nursery rhyme 'Ring a ring of roses, a pocketful of posies, atishoo, atishoo, we all fall down.' The posies referred to the flowers that terrified people used to carry near their noses, in the vain hope of warding off the plague's 'foul vapours'.

The third strain of the plague is septicaemic plague, which is carried by insects and is almost always 100 per cent fatal. Luckily, it is extremely rare.

What is DNA?

O ne of the things that puzzled and no doubt frustrated the British naturalist Charles Darwin (1809-82) was that, while working on his book entitled *On the Origin of the Species by Means of Natural Selection*, describing his theory of evolution, he was unable to explain the process of inheritance: in other words, how offspring inherited the characteristics of their parents. Yet the key to the puzzle was already in place, thanks to the pioneering work of the Austrian biologist Gregor Mendel (1822-84), who suggested that all characteristics in sexually reproducing species are inherited through what he called 'factors' contributed by each parent to its offspring. These 'factors' were later called genes, and through the efforts of many scientists it is known that genes are part of the chromosomes located in the nucleus of a living cell.

In 1944, Oswald T. Avery, Colin MacLeod and Maclyn McCarty, all New York scientists, proved beyond all doubt that the hereditary-transmitting substance was deoxyribonucleic acid, abbreviated to DNA, which is present in all cells together with ribonucleic acid (RNA), the latter being involved in transforming DNA into proteins.

It was not until 1953 that the structure of DNA was unravelled. The British molecular biologist Francis Crick and the American biochemist James Watson, an ornithogolist by training, in an article in the April 1953 edition of the journal *Nature*, described the correct structure of the complex giant DNA molecule – a molecule that contains, in coded form, all the information that is needed to build, control and maintain a living organism.

The work of the two scientists was considerably aided by earlier research carried out by the Anglo-New Zealand physicist Maurice Wilkins, who, together with his British colleague Rosalind Franklin, had been using X-ray crystallography techniques to examine the makeup of atoms in a molecule. Crick and Watson applied this technique to the examination of the large and complex molecules found in living cells, and were able to show that DNA had a double helical structure, reminiscent of a spiral staircase.

The DNA molecule is made up of two chains of nucleotide (organic compound) sub-units, with each nucleotide containing either a purine (adenine or guanine) or pyrimidine (cytosine or thymine) base. The bases link up with each other – the adenine linking with thymine and the cytosine with guanine – to form base pairs, connecting the two strands of the DNA molecule to form the double helix. The way in which the pairs form ensures that the base sequence is preserved from one generation to another.

Sets of three bases are known as codons; it is here that hereditary information is stored, the blueprint for the manufacture of a particular amino acid, the sub-unit of a protein molecule. The information stored in each codon is transcribed by the messenger molecules of RNA and is translated into amino acids in the ribosomes and cytoplasm, the protein-making machineries of the cell. The sequence of codons determines the exact order in which amino acids are linked up during manufacture, and therefore the kind of protein that is to be produced. Because proteins are the building-blocks of all living matter and are also enzymes, which regulate every

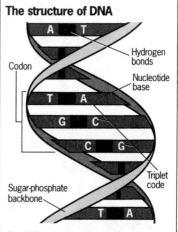

The structure of DNA

Codon
Hydrogen bonds
Nucleotide base
Triplet code
Sugar-phosphate backbone

The DNA molecule consists of two strands wrapped around each other in a helix. The main strands consist of alternate sugar and phosphate groups, and attached to each sugar is a nucleotide base: adenine (A), cytosine (C), guanine (G) or thymine (T).

aspect of metabolism, the genetic code is responsible for building and controlling the whole of a living organism.

The DNA molecules in a cell nucleus contain up to ten million atoms in two intertwined spirals, and when a nucleus divides, the DNA molecule has to spin round at great speed – several hundred revolutions a second – to unravel them. The reading and copying of a genetic message happens in an incredibly short time, proteins being assembled from amino acids in seconds.

The unravelling of the DNA puzzle – which brought Crick, Watson and Wilkins the award of the Nobel Prize for physiology or medicine in 1962 – laid the foundations of genetic engineering, which is revolutionizing biology and medicine today.

What was the earliest form of life?

When scientists began to study the origins of the Earth, they divided its history conveniently into three eras, naming each according to the status of life on the planet at that time as it was revealed by the fossil record. Going back in time from the present, these eras were the Cenozoic, meaning recent life; the Mesozoic, middle life; the Paleozoic, early life; and the Precambrian, which was also known as the Azoic – without trace of life.

They were wrong about the Precambrian. Although this era furnishes the least information of all about Earth's history, it extends across nearly 90 per cent of it, stretching from 4.5 billion years ago, when the planet condensed from the cloud of gas and debris circling the new-born Sun, to 570 million years ago, when the rocks reveal the first evidence of living cells combined in the form of multicellular organisms.

Because we cannot see the microcosm with the unaided eye, we tend to dismiss its significance. Yet of the three and a half billion years that life has existed on Earth, the evolution of human beings from their early primate ancestors to the present day represents far less than one per cent. We tend to overlook the fact that, for the first two billion years, Earth was inhabited by bacterial microorganisms, each barely a thousandth of a millimetre in diameter, teeming through the warm, soup-like oceans of the nascent world, the products of evolution from the melting-pot of raw chemicals that formed the basis of creation.

These microorganisms are called prokaryotes, from the Greek *pro*, meaning before, and *karyos*, meaning kernel or nucleus. They are the smallest and simplest of all living things, single-cell organisms with a simple DNA molecule floating free within the cell. They are also the most widely dispersed of all living things, inhabiting every kind of environment from the ocean depths to the vents of volcanoes, and from the polar icecaps to the near-boiling water of natural hot springs. They can remain frozen, or desiccated, or otherwise dormant for years and then return to life.

In their first two billion years, prokaryotes continuously transformed the Earth's surface and atmosphere, as well as invented all of life's essential, miniaturized chemical systems. Their activity led to the development of fermentation, photosynthesis, oxygen breathing and the elimination of nitrógen gas from the air. It also led to widespread pollution that resulted in the mass extinction of other microorganic life forms at the dawn of time.

Over hundreds of millions of years, prokaryotes mutated and adapted to fit in with a changing environment. Somewhere along the line – and the mutations must have happened countless times – the process produced an organism that was able to harness the radiant energy of the Sun to sustain it, an innovation that has directed the course of life on Earth ever since. Instead of using the chemicals from its surroundings to provide growth and power, the organism used sunlight, making it far more energy-efficient than anything that had gone before. This was the earliest form of photosynthesis.

Photosynthesis needs a source of hydrogen, and about three billion years ago an organism developed that was able to split water molecules into their component parts: two parts hydrogen and one part oxygen. The hydrogen was used in the photosynthetic process; the oxygen was released into the atmosphere in ever-increasing quantities, and some groups of prokaryotes adapted to it and even became dependent on it, as are all animal life forms today.

Some, however, did not. These were the methanogens, bacteria which are found in marshes, lake beds and the digestive tracts of animals. Oxygen is poisonous to them. They can live off inorganic chemicals without the aid of any other energy source. In the process they discharge methane, the marsh gas that ignites in contact with oxygen to produce the will-o'-the-wisp.

They are here now; they were here in the beginning.

What is an aura?

Throughout the long history of the world's mystical and spiritual traditions, it has been claimed that people are surrounded by a region of hazy light that can be detected by people with psychic sensitivity. This aura reflects the state, wellbeing and spiritual development of a person by its colour and degree of strength. In art, it is depicted by the golden halo that surrounds the heads of especially holy people. But does such a detectable force field really exist?

The first scientific evidence that it might came in the 1890s, when a Polish scientist called Jacob Nerkiewich, working in Paris, photographed what appeared to be a luminous glow around the body of a person immersed in the powerful, high-frequency magnetic field of a high-tension electrical transformer. Then, in 1939, two Russian engineers, Semyon and Valentina Kirlian, began a series of experiments that involved taking photographs of living objects, from human hands to the leaves of trees, in a powerful electric field. They discovered that, around the edges of whatever organic and living object they photographed in these conditions, there was what seemed to be an aura. By the 1960s, the Kirlians had developed a reliable method of photographing an electrical discharge around the human body, a technique that came to be known as Kirlian Photography.

The method involves placing a film on an insulating pad, which is in contact with a metal plate connected to a source of high tension electric current. An organic specimen is placed on the film and the circuit is switched on. The film records the pattern of ionization that takes place in the insulating material due to the interaction of the high-tension field and the electromagnetic field of the specimen.

Kirlian Photography has proved a useful method of exploring the normal and the diseased pattern of living things, and some argue that it appears to prove that living organisms do indeed possess auras; it is noteworthy that it does not work with inorganic objects, such as stones, although an aura has been shown to persist for some time after the death of a living object. A classic example of this is the so-called phantom leaf effect, where a leaf with a piece torn off is placed on a Kirlian photo plate and the high-tension source turned on. What appears on the resulting photograph is an image of the whole leaf.

The idea that every living organism possesses a 'life force' that extends for a short distance beyond its physical body is not really far-fetched. After all, living things produce a whole range of electromagnetic energy as part of their life-sustaining processes. This energy is well known in traditional Chinese medicine, where it is called *Chi*; it is present everywhere, in the body and in the earth, where it follows paths known as 'dragon lines', and it also forms the basis of acupuncture techniques. The Hindus call it *Prana*, the breath of Brahma.

It appears, then, that around the physical bodies of living organisms there is a region of ionized air produced by the organism's own electromagnetic field, and that under certain circumstances this region can become visibly luminous, in much the same way that regions of the upper atmosphere become luminous under the ionizing effect of the solar wind to produce an aurora. It would also appear that the effect is visible only to certain people who are attuned to the appropriate wavelengths, although there is reason to suppose that it is visible – or at least sensible – to animals, whose senses can detect wavelengths in the electromagnetic spectrum that the senses of human beings cannot.

What is telepathy?

Many of us have experienced the phenomenon we call telepathy at some time. It may have come in the form of a sudden mental shock, an indication quite out of the blue that something was wrong with someone close to us; or it may have occurred in a dream, where we received a message that someone was trying to contact us urgently – which, on investigation after waking, turned out to be true.

A popular dictionary defines telepathy as 'communication from one mind to another at a distance other than through known senses'. You will not find the word in a specialist scientific dictionary, though, because orthodox science will not admit that any such thing exists.

At the height of the cold war, however, both the Americans and Russians were sufficiently interested in telepathy to carry out some serious experimentation in schools, colleges and miltary establishments. In 1959, for example, the nuclear submarine USS *Nautilus* spent 16 days under the Atlantic while a scientist on board concentrated on sending thought-pictures of various symbols to a colleague 1000 miles away in Maryland, and in the early 1960s more experiments of this kind were carried out by scientists on two more American submarines, the *Skate* and the *Skipjack*. Then, in 1967, the Russians revealed in a scientific magazine that they had been carrying out a comparable series of tests with the aid of one of their own nuclear submarines, the *Vityaz*. Experiments in telepathy were also carried out between ground stations and astronauts in orbit.

The purpose of the tests, which were described as 75 per cent successful by the Americans, was twofold. Firstly, there was an obvious application for telepathy in the world of espionage; imagine the implications of an agent being able to scan the pages of a secret document and send their contents back to his headquarters by thought transference, instead of having to resort to high-risk copying and communication techniques. Equally as important, if not more so, was the possible application of telepathy to underwater warfare. The experiments were undertaken at a time when the USA and USSR were building large fleets of missile-carrying submarines, capable of lurking in the ocean depths or under the Arctic ice cap for months on end with barely a chance of detection; it was the ultimate deterrent, but it had an attendant problem of communication.

Water acts as a barrier that strongly absorbs all electromagnetic waves except blue-green light and extremely low frequency (ELF) waves. During most of the cold war period very low frequency (VLF) waves were used to communicate with submarines, but these penetrate only a short distance into the water, so a submarine had either to approach the surface or send up antennae to receive signals at known transmission times, thereby increasing its vulnerability. If it were proven that telepathic waves could penetrate deep layers of sea, the communications problem would have been overcome.

Unfortunately, it never was proven, although some researchers continue to claim that the ability to develop telepathic powers is latent in the minds of all of us, and that all it needs is the discovery of the right electromagnetic wavelength – which we may accidentally hit upon in moments of distress in a subconscious attempt to communicate with those close to us.

What is antimatter?

A t 7.17 in the morning of 30 June 1908, passengers on board the Trans-Siberian Express east of Krasnoyarsk were startled by a roar overhead. Looking up, they saw a great, glowing object race across the sky from south-west to north-east. A few moments later, the northern sky burst asunder in an intense glare of light. Minutes passed, and then the train was shaken by a colossal shock-wave. In the north-east, beyond the horizon, a pillar of smoke mushroomed skywards.

The passengers on the express would have been even more startled and terrified had they known that the brilliant flash, and the epicentre of the smoke cloud, were more than 350 miles (560 km) away. At Keshma, 120 miles (190 km) from the explosion, the burst of light was so strong that it outshone the Sun and cast its own shadows, while the blast hurled men, horses and cattle to the ground.

Perhaps the most graphic eye-witness account of the event was provided by a farmer named S.B. Semyonov, who lived beside the Tunguska River, a tributary of the Yenesei, some 40 miles (60 km) from the blast. He had just left his home when the tremendous light flashed across the sky.

'I scarcely had time to shield my eyes with my hand before the flash of light died away. It grew dark, and the explosion came a few minutes later. The blast blew me back into my house and threw me to the ground. The heat was so fierce it scorched my clothing. The ground rose and fell like a wave on the sea.'

Even today, it is not clear whether there were people who experienced the explosion from even closer quarters, and lived to describe it. There were rumours that two villages had been wiped out.

The shock-wave from the explosion in central Siberia was registered on seismographs all over the world. In addition, the sky over the whole of Europe shone with a strange luminous glow for two nights afterwards, and the Sun rose and set in a rainbow of vivid colour.

Exactly what it was that exploded in the Tunguska area of Siberia in 1908, causing enormous damage to the forest for miles around, has been the subject of controversy ever since, with theories ranging from asteroids, to comets, to alien spacecraft. But one more recent theory is interesting, because it touches on a subject at the frontiers of science. The subject is antimatter.

In 1928, the British physicist Paul Dirac suggested that there should be an analogue of matter called antimatter. Taking the important subatomic particle, the electron, as an example, he worked out mathematical equations to show that the electron – which has a negative electrical charge – must have a counterpart with a positive charge. He was right; in 1932 just such an 'antielectron' was discovered by the US physicist Carl Anderson. The new particle had the same magnitude of mass and charge as the electron, but possessed a positive charge. It was named the positron.

Today, we know that for every type of particle, there is a corresponding antiparticle, and in the same way that matter is constituted of fundamental particles, antiparticles are the building blocks of antimatter. They can be created in particle accelerators.

The problem with matter and antimatter is that when they come into contact, they annihilate one another in a stupendous burst of energy. According to some theories, the whole universe is filled with so-called virtual particles, which come into existence spontaneously as particles of matter and antimatter and immediately cancel one another out. But suppose some antimatter emerged and was not cancelled out, and suppose instead it was created in Earth's atmosphere? The result would be a cataclysmic explosion, and the

Tunguska blast was certainly that – the equivalent of about 20 megatons.

It remains just a theory, of course. But in the absence of more concrete evidence, it's as good as any other.

What is gout?

He was a favourite target for Victorian cartoonists – the gouty, florid-faced country squire, foot swathed in bandages, waving his stick irritably at luckless servants.

The squire couldn't really be blamed for feeling irritable. As anyone who has suffered from it knows, gout is no laughing matter. And, although there are drugs to help alleviate the suffering it causes, it isn't something that can be wiped out overnight.

Gout is a metabolic disorder that causes severe attacks of arthritis. An acute attack usually affects only one joint, mainly the one at the base of the big toe, but it can also attack other joints, including the knee, ankle, wrist, foot, and the small joints of the hand. The affected joint becomes red, swollen, and extremely tender. Sometimes, the redness spreads over a considerable area and may be confused with other conditions, such as inflammation of the tissue under the skin. The swelling is accompanied by such intense pain that the sufferer may not be able to place the affected foot on the ground, or even tolerate the pressure of bandages or bedclothes on it. The pain, which sometimes goes hand-in-hand with a mild fever, reaches its peak within 24 to 36 hours.

The first attack usually lasts only a few days. Some people never experience the discomfort again, but most have a second attack between six months and two years after the first. Subsequent attacks may affect more joints, and severe inflammation may lead to permanent damage and constant pain.

The Victorians attributed attacks of gout to rich food and high living – hence the lampoon of the suffering squire, stuffed with game and wine. In fact, anyone can fall victim to it. Martin Luther did, and he was a pretty austere character. There is a genetic element in 75 per cent of cases and gout may also be associated with other complaints such as kidney stones. However, the basic cause of gout is a high level of uric acid in the blood and other body fluids. An abnormally high level causes crystals of monosodium urate monohydrate to be deposited in the joints or adjacent tissues, and it is this that leads to the swelling and the pain. Gout affects ten times more men than women; in men it occurs at any time after puberty, but in women it generally only occurs after the menopause.

Although the increase in uric acid in the body may be caused by hereditary factors or by disease, the most usual cause is a rise in the metabolism of purine, a nitrogen-rich compound that is produced by the digestion of certain proteins. Foods high in purine include sardines, liver, kidneys, pulses and poultry, and should be avoided by people who suffer regularly from the complaint. Luckily, it is not necessary nowadays to stick to a purine-free diet. The pain and inflammation caused by gout can be controlled with non-steroidal anti-inflammatory drugs, or if these are not considered suitable, by colchicine, an extract of the autumn crocus, which was first used late in the 18th century.

The drug allopurinol is also widely used in the long-term treatment of gout. It does not alleviate the pain caused by acute attacks, but it reduces their frequency over a period of time. However, in the first few weeks of treatment it may sometimes increase the frequency of attacks, and in some cases may produce side effects such as itching, rashes and nausea.

Even though gout may be confused with other ailments at first sight, it is not difficult to diagnose. Fluid is removed from the affected joint and examined under a microscope, and the presence of uric acid crystals confirms the diagnosis.

There may have been some truth in the old belief that gout was a rich man's complaint. Up to the Second World War, there was a great deal of difference between the diets of rich and poor; poultry, for example, was a luxury that graced the tables of the less wealthy only at Christmas. Today, the type of food we eat is much more common to all levels of society, which has made gout less selective.

One indulgence that may trigger off an attack is heavy drinking, particularly on an empty stomach. But then, that can lead to all sorts of miseries.

What is a pearl?

After diamonds and emeralds, no gemstones are more prized than pearls – and yet they are nothing more than the products of molluscs with indigestion.

A pearl is a shiny, round, hard abnormal growth consisting of nacre, more popularly known as mother-of-pearl. Nacre is composed of aragonite, which is one of the components of a mollusc's shell; the main constituent is a hornlike substance called albuminoid. The cells that secrete the substances that make up the shell are located in the mantle, a fold of skin that covers all or part of the body. When a foreign particle – a parasite, grain of sand or some other irritant body – penetrates the mollusc, cells attach themselves to the particle and build up more or less concentric layers of nacre around it, and over a period of time, usually several years, a pearl is formed. If the nacre builds up around muscular tissue it often has an irregular shape and is known as a baroque pearl; blister pearls are the result of nacre building up adjacent to the shell, and are usually flat on one side.

Pearls come in a variety of delicate shades; they can be black, white, grey, blue, yellow, green, mauve, lavender or rose. The colour depends on the type of mollusc and its environment, with roseate pearls being among the most prized of all. Although commercially valuable pearls are obtained from both freshwater mussels and oysters, the most precious examples are produced by the pearl oysters of the family *Pteridae* whose various species live in the Indian Ocean, where roseate pearls come from, the tropical waters off northern and western Australia, off the coast of California and in the Persian Gulf. Some of the world's finest pearls are to be found here, the produce of a saltwater mollusc called the mohar, which lives at depths of between eight and 20 fathoms (48 to 120 ft/15 to 37 m).

Freshwater mussels can produce pearls of perfect shape, which makes them worth more than pearls produced by saltwater oysters. The Mississippi River is a prime source of choice freshwater pearls, as are the forest rivers of central Bavaria and other locations in central Europe. Freshwater pearl fishing has also been an industry in China since before 1000 BC.

The weight of pearls is measured in grains; one pearl grain is equal to 50 milligrams, which in turn is equal to ¼ carat. Pearls weighing less than ¼ grain are called seed pearls. The largest pearls occurring naturally are baroque pearls; one such is known to have weighed 1860 grains.

In 1893, the Japanese discovered how to cultivate pearls

artificially, by inserting a tiny bead of shell from a clam, together with a small piece of membrane from the mantle of another pearl oyster, into oysters kept in cages in the sea. After three years, these 'cultured' pearls are ready for harvesting. Today, the cultured pearl industry is widespread throughout the world, and this has led to a decline in natural pearl fishing.

What is silk?

Apart from the bee, there is another insect that has become almost completely domesticated through rendering its services to mankind over the centuries. It is a moth caterpillar called *Bombyx mori*, more popularly known as the silkworm.

The silkworm, a creamy-grey caterpillar about 3 in (8 cm) long when fully grown, has been domesticated for at least 4000 years and is an amazingly industrious worker. In three days of constant weaving it spins a cocoon from which can be drawn a filament between 2000 to 3000 ft (600 to 900 m) long and only 1/1200th of an inch thick. About 25,000 cocoons make a pound of raw silk.

Cultivation and care of the silkworm caterpillar continues from the egg stage until the cocoon is completed, and runs parallel to the cultivation of the mulberry trees on which the larvae feed. To build its cocoon, the silkworm weaves a long, continuous fibre around itself. As it does so, it excretes liquid secretions from two glands via its spinneret, a single tube in the head; these harden when they come into contact with air to form twin filaments made up of a protein material called fibroin. A second pair of glands produce a gummy substance called sericin, which binds the filaments together. Since an emerging moth would break the cocoon and ruin the whole process, steam or hot air is used to kill the larva as it enters the chrysalis stage, a necessary if rather sad ending for a creature that has performed so valiant a task.

The silk is freed from the cocoon by softening the sericin 'cement'. The ends of the filaments from several cocoons are located and the filaments from them drawn out at the same time to form a single strand, several strands being twisted together to make them thicker and suitable for commercial use. At this stage the silk still contains the gummy sericin substance and is known as raw silk; sericin gives protection during processing, and is removed at the yarn stage by boiling the silk in soapy water. This reduces the weight by as much as 30 per cent, but leaves the product soft and lustrous.

The thickness of silk filament yarn is expressed in terms of denier, which is the number of grams of weight per 9800 yd (9000 m) of length. Spun silk, which is obtained from short lengths broken off during processing or from damaged cocoons and twisted together to make yarn, is expressed in terms of the number of hanks per pound, each hank being 840 yd (768 m) in length.

Silk is very strong, with a breaking strain of over 0.1 oz (4 g) per denier. It can be stretched about 20 per cent beyond its original length before breaking. It is lower in density than cotton, wool or rayon and has excellent moisture-absorbing properties, retaining as much as a third of its weight in moisture without feeling damp. It can also resist temperatures of up to 340°F (170°C), which makes it more heat-resistant than wool. The notion that the rustling noise made by silken garments is an indication of quality is a myth; it is a result of the processing treatment.

It was the requirement for silk that opened up the ancient trade routes between China and the West, the 4000-mile (6400 km) silk road beginning at Sian and ending in the Levant. After the collapse of the Roman Empire in the east it became unsafe, but it was revived under the Mongols and later used by the Polo brothers on their celebrated expedition to Cathay.

Apart from its colour – it readily accepts most dyes – and its lustre, silk was much prized in the mediaeval western world for two reasons. Firstly, it did not soil easily, since when it was subjected to friction it acquired a charge of static electricity, especially in low humidity, and secondly, it did not harbour the constant companion of mediaeval men and women – the body louse.

What is coral?

The word coral really applies to two things: certain types of marine invertebrates and the skeletons they leave behind to form characteristic reefs.

The marine creatures belong to the class Anthozoa (*phylum Cnidaria*) whose members have external or internal skeletons of a stonelike, horny or leathery consistency. They include sea anemones and jellyfish. The skeleton consists of lime (calcium carbonate) extracted from the surrounding water. Corals live in symbiosis with microscopic algae which obtain carbon dioxide from the coral polyps; in turn, the polyps receive nutrients from the algae. The polyp is the coral animal's body, a hollow, cylindrical structure attached at its lower end to some surface. The mouth is at the free end and is surrounded by tentacles which gather food; they are equipped with stinging structures called nematocysts, which they use to paralyse their prey. Eggs are also ejected through the mouth and the larva, known as a planula, swims freely for several days or even weeks before settling on to a solid surface and becoming a polyp. The polyps live in colonies, and as new ones develop the old ones underneath die, leaving their skeletons.

Stony corals are the most widely distributed forms. They are present in all the world's oceans from the tidal zone to depths of nearly 20,000 ft (6000 m). Their polyps range in diameter from 0.04 to 1.2 in (1 to 30 mm). In colour they are olive, yellowish or brownish, depending on the colour of the algae living on them, but the skeletons they leave behind are white and are composed of almost pure calcium carbonate.

Atolls and coral reefs are composed of stony coral, these formations growing at an average rate of 0.2 to 1.1 in (0.5 to 2.8 cm) a year. The Great Barrier Reef, to the north-east of Australia, is about 1000 miles (1600 km) long, has a total area of 7700 sq miles (20,000 km^2), and increases by 50 million tonnes of calcium every year as the tiny skeletons build up.

Worldwide, coral reefs cover an estimated 240,000 sq miles (620,000 km^2). Barrier reefs are separated from the shore by a salt water lagoon, which may be as much as 20 miles (30 km) wide, and there are usually breaks in the barrier permitting access to the lagoon, which is why these areas are often happy hunting grounds for sharks. Often, coral builds up on the shores of continents or islands, producing so-called fringing reefs.

Coral atolls are different in that they are formed by the subsidence of an extinct volcano, and are built up on the perimeter of the sunken volcanic island.

What is a jet engine?

The answer to this might sound pretty obvious. After all, millions of people are borne around the airways of the world by jet engines every year, on holiday or on business. They are a necessary part of modern life. And yet, how many people who routinely travel by air have the faintest idea of what goes on inside a jet engine, and of how the principle works?

Like a rocket motor, a jet engine is a reaction engine. It obeys Isaac Newton's third law of motion, which states that to every action there is an equal and opposite reaction. In the case of jet and rocket engines, an object – the engine and also the airframe that is built around it – is propelled in one direction by a stream of gases moving in the other. The main technical difference between the jet and the rocket is that the latter does not need to draw on oxygen from the atmosphere as part of the combustion process; it carries its own oxidizer internally in the form of liquid oxygen, or else uses a solid-fuel propellant.

All the jet engines in the world today stem from the pioneering work of Britain's Frank Whittle and Germany's Hans von Ohain, who pursued the concept independently in the years before the Second World War. The generation of early jet engines resulting from their research were simple gas turbines, in which air passes through a forward-facing intake, is compressed by a compressor fan, and is then fed into a combustion chamber where it is mixed with an atomized spray of fuel – usually kerosene – and ignited. The resultant hot gases expand rapidly and exit from the rear of the combustion chamber, spinning a turbine that drives the compressor via a driveshaft. The initial spin of the compressor, necessary to draw in air through the intake, is imparted by either an electrical source from an external battery, called a trolley-accumulator (trolley-acc, in familiar aviation terms) or an explosive cartridge inserted into the engine casing. Gas turbines, or turbojets, are used to power military aircraft well into the supersonic speed range; the thrust they deliver is proportional to the mass of gas they produce multiplied by the acceleration imparted to it. It is expressed in either pounds or kilograms of static thrust or force, both of which are now being superseded by the international unit, the Newton (one Newton is the force needed to accelerate an object with a mass of one kilogram by one metre per second).

A later development, the turbofan engine, is fitted with an extra compressor. Some of the air by-passes the combustion chambers and mixes with the jet exhaust, reducing its temperature and velocity. The overall result is greater efficiency, economy and quietness at high subsonic speeds, and it is this type of engine that is widely used in airliners. Another type of engine, the turboprop, is best suited for operation at lower speeds and altitudes; this derives its thrust partly from the jet of exhaust gases, but mainly from a propeller powered by a turbine that is driven by the jet exhaust.

The simplest form of jet engine is the ramjet, in which air is literally rammed into the engine without the need for a compressor. The problem with this type of engine is that it has to be accelerated to high forward speed before it can function, so in its initial flight phase it has to be boosted by some means, usually a rocket motor. It has been widely used in long-range surface-to-air missiles. A form of ramjet, the pulse jet, in which air is admitted in 'pulses' by means of the rapid opening and closing of a shutter system at the intake, was used in the wartime German V-1 flying bomb, which had to be booster-launched from a ramp or air-launched to achieve the necessary forward speed.

Since the 1950s, most military jet engines have been fitted with reheat, or afterburning, in which fuel injected into the hot exhaust stream provides a thrust increase of up to 70 per cent for short periods. It is fuel-consuming and is therefore normally used only for combat manoeuvring or to assist take-off at high all-up weights.

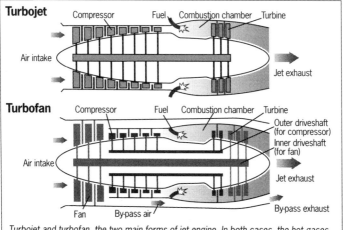

Turbojet and turbofan, the two main forms of jet engine. In both cases, the hot gases produced by the combustion process are expelled at high speed from the rear of the engine, imparting forward motion and turning a turbine which drives the compressor. In the turbofan, some of the airflow bypasses the core engine and mixes with the exhaust stream, giving lower temperature and velocity – and greater economy.

One of the most promising jet engine developments is vectored thrust, in which the jet nozzles can be swivelled through 90 degrees from downward vertical to rearward horizontal to provide vertical and short take-off and landing capabilities. It is used in the British Aerospace/McDonnell Douglas Harrier/AV-8 strike aircraft. Thrust vectoring in the form of reverse thrust is used to slow down aircraft on landing; the jet pipe is blocked by special baffles and the exhaust gases redirected forwards through cascades in the engine nacelle.

What is a mirage?

We have all heard stories of desert travellers being lured to their doom by a mirage – the sudden appearance on the horizon of, say, signs of human habitation or an inviting pool of water.

A mirage is an optical illusion caused by the refraction, or bending, of light rays as they pass through atmospheric layers of varying density. Under appropriate conditions, such as when an expanse of desert is heated by intense sunshine, the air cools rapidly as it ascends, and as it does so it increases in density and refractive power.

Some of the sunlight that is reflected downward from the upper part of an object – a tower, for example – travels through the cool air in the normal way in a straight line and reaches the eye of an observer without being refracted, so that a direct image of the tower is seen; a portion of the light, however, enters the hot and rarefied air near the ground and travels in a downward curve before bending upwards, so that when it reaches the observer's eye it produces the illusion that it originated beneath the heated surface. This image is inverted, so the observer appears to see the tower with its image reflected in water. Because the light is from a blue sky, the mirage effect can make the horizon itself appear blue and watery, giving the illusion of a lake, or an expanse of sea.

In cooler conditions, such as during the desert night or over a body of water, the effect is exactly the opposite. In this case, the cooler air collects in a layer near the ground, with the warmer air in a layer above it, so that light rays may be deflected upwards in a curve before descending to reach the eye. As a consequence, an observer may see objects that would normally be out of the line of sight, like a ship below the horizon, apparently floating in the sky. The phenomenon is called 'looming'.

What is amber?

Coloured in every shade of yellow, often with hints of orange, brown and sometimes red, amber is one of the world's natural treasures. It is found throughout the world, but the largest deposits occur on the shores of the Baltic Sea, where the sands are anything up to 60 million years old.

Amber is fossilized tree resin which, having lost its volatile constituents and undergone various chemical changes, has reached a stable state. Often, pieces of it contain many minute air bubbles; these occur particularly in opaque varieties, which are called bone amber.

Modern scientific techniques, including the use of infrared spectroscopy, have shown that certain varieties of amber millions of years old are related to modern resin-producing trees.

Commercially, amber is used to make carved objects such as beads, rosaries, cigarette holders and pipe mouthpieces. Amberoid is a synthetic form which is made by fusing together fragments of amber under pressure; it contains parallel bands, which distinguish it from the real thing.

The real value of amber, however, is in its preservation of the bodies of insects that lived during the Middle Tertiary period, the era of geological time that extended from 65 million to 1.64 million years ago. Scientists are hopeful of extracting DNA from some of the best-preserved specimens, and even re-creating them under laboratory conditions.

Sometimes, however, insect specimens trapped in amber are not all they seem to be. In 1994, the British Natural History Museum received a severe shaking with the revelation that one of its most treasured fossils, a fly trapped in a 38-million-year-old piece of amber, was nothing of the sort. It turned out that the fly was creeping around in a German sewer barely a century ago. It fooled generations of scientists until a graduate student, taking part in an experiment to extract DNA, discovered that the fly was a very clever fake.

Using a powerful microscope, he found that the amber had been cut in half, the fly embedded in resin inside a small hollow, and the two halves glued together again. The fake fossil, which first turned up in a collection started in 1850 by a German fly expert, had been sold to the Natural History Museum in 1922.

What is tuberculosis?

In Queen Victoria's time, the disease tuberculosis (TB) was a scourge that killed one in four people in western Europe. It remained a serious health hazard up to the Second World War, when the advent of effective antibiotics caused it to decline.

But the decline was not worldwide. A third of the world's population is still infected with TB and the disease continues to kill three million people every year, 95 per cent of them in developing third world countries.

In humans, tuberculosis is caused by the bacterium *Mycobacterium tuberculosis*. Many people infected with it never develop the active disease. The chances of this happening are about ten per cent, but the risk increases greatly if a person's immune system is affected by other causes such as HIV, malnutrition – or the march of old age.

Tuberculosis is easily spread. The bacterium travels in airborne droplets, so that it can be spread by coughing, sneezing or even talking. However, provided there is nothing wrong with the immune system, the risk of infection through chance contact is relatively low. The danger is further reduced by good ventilation in households and public places.

TB's renewed onslaught on the western world became seriously apparent in New York in the 1980s, where cases of infection leaped by nearly 150 per cent. Worst hit were the city's poorer quarters, where the problem was complicated by drug abuse, inadequate diet and the debilitating effects of other diseases.

The most alarming factor is that the TB bacterium is becoming drug-resistant. Some strains have even developed resistance to all known and tested drugs, and it is virtually impossible to treat people infected with these strains.

The cheapest and most widely used vaccine against TB is *Bacille Calmette Guerin*, or BCG for short, which is a live bacterium extracted from cattle. Over the years it has saved millions of lives, and most schoolchildren in Britain are now treated with it as a preventive measure. But although BCG works well with children, it produces mixed results when given to adults, sometimes providing no protection at all.

One major problem in fighting TB today is that it can take up to a month for hospital tests to confirm if a patient is infected, and even longer to determine whether a person is infected with a drug-resistant strain of the bacterium. Even if the process of identifying the TB bacterium in patients can be speeded up, medical science is up against the fact that no new anti-TB drug has been developed for years. Research into TB stopped when scientists thought that the disease had been checked, and that was before the modern era of biotechnology.

This is changing. The emergence of drug-resistant TB strains has given impetus to renewed research, particularly into links between the disease and HIV. Immune deficiency leads to a much increased susceptibility to TB, as to other infections, and is very common in people with AIDS.

What is a jet stream?

To understand this, we must first of all understand how winds circulate around the globe.

When warm air rises near the equator, it cools and moves outwards at high altitudes before descending again in the sub-tropical regions on either side of the equator at about 30 degrees of latitude. At the same time, low-altitude winds return air towards the equator to replace the rising air, creating a circulation pattern on either side. This is known as a Hadley cell after George Hadley, an 18th-century British meteorological pioneer.

Outboard of these two circulation patterns, warm air is transferred towards the poles and in turn is replaced by warm air moving towards the equator. In general, rising air is associated with low pressure at the surface, and produces rain as it rises; descending air, on the other hand, is associated with high pressure at the surface, and is dry. All this interaction, presented in a greatly simplified form, explains why the tropics are warm and wet, and why they are bordered by arid regions.

The next factor to be considered is the Earth's rotational speed, which is very low at a point near the poles but just over 1000 mph (1600 km/h) at the equator, reducing to 500 mph (800 km/h) at latitude 60 degrees. The rotation, of course, is from west to east. Air that moves from the equatorial regions towards the poles carries a momentum of this high rotational speed, and when it returns to the surface at higher latitudes it is still moving relatively fast. This air, deflected eastwards by the forces acting upon it, forms the prevailing winds that occur in both hemispheres between latitudes 35 and 60 degrees: the Westerlies. And it is in the Westerlies that jet streams are born.

Jet streams are the product of pressure and temperature differences between sub-tropical high pressure cells and areas of lower pressure towards the poles, and of the related temperature and energy transfer. They occur at altitudes of between 30,000 and 50,000 ft (9,000 and 15,000 m) in the upper troposphere and lower stratosphere; on average their speed is 150 mph (240 km/h), but at times, especially in the winter, they can reach speeds of 300 mph (480 km/h). In effect they are fast-moving masses of laterally-concentrated air.

In the northern hemisphere there are two main westerly jet streams, the more northerly of which is the Polar front jet stream. This is associated with the steep temperature gradient where polar and tropical air interact. The other, the more persistent of the two, is the subtropical jet stream, which is related to a temperature gradient confined to the upper atmosphere. In 1976, this jet stream became blocked by a region of high pressure centred on the British Isles and divided around it, becoming a contributory factor in the lengthy period of drought experienced in north-west Europe that summer.

In the summer, an eastern tropical jet stream forms in the upper troposphere over India and Africa because of a regional reversal of the south-north temperature gradient; this summer pattern of circulation accompanies the final burst of the monsoon with the arrival of the humid, low-level south-westerlies. A further jet stream, known as the Somali jet, is at its most active during the Indian monsoon and flows north-westward from Madagascar over East Africa. This stream reaches a speed of up to 100 mph (160 km/h) and forms a key part of the south-west monsoonal flow, but unlike other jet streams it is active at a level of only 3250-5000 ft (1000-1500 m.)

Jet streams, then, are a key factor in the Earth's complex pattern of climate. Intense forces are produced in them, but even so man has learned to harness their power. The velocity of a jet stream can contribute greatly to an aircraft's tailwind component, increasing its ground speed, reducing flight time and therefore producing greater economy.

What is Damascus steel?

The heavily armed Crusaders must have thought they had it all their own way when they launched their first attempts to recapture the Holy Land. Their Muslim enemies, the Saracens, wore only light armour, and they had to face the pick of Europe's renowned heavy cavalry.

Yet the Crusaders soon found, to their cost, that the Saracens had a secret weapon. It was a sword forged of steel so durable that it would not shatter; it was light, and yet sharp enough to slice through armour. Before long, the name Damascus Steel had become legendary throughout the kingdoms of mediaeval Europe, and captured Damascus swords were prized heirlooms handed down from father to son.

The Crusaders learned that the steel for the Damascus blades was made in India, forged in Persia, and the weapons sold on the arms market in Damascus, Syria. Some of the Crusaders managed to get hold of unfinished steel blocks and bring them home – and it was here the mystery began. No European swordsmith, however skilled, could produce a perfect Damascus blade.

The problem was that when they hit the white-hot steel with a hammer, it shattered. Moreover, the swordsmiths were confused by the tales the Crusaders had brought back with them: tales probably spread deliberately by the weapons manufacturers in Persia to preserve their secret. One story was that the blade had to be tempered by plunging it into the body of a slave; another, that it had to be soaked in urine.

In the end the swordsmiths gave up, and over the centuries the secret of Damascus steel was lost – until it was rediscovered quite recently and accidentally, by metallurgists working in the aviation industry.

They were investigating the properties of what is called superplasticity in metals – the point at which metal, when heated to a certain temperature, becomes elastic enough to be stretched and shaped without breaking, and then keeps its original strength after it is cooled. The reason for this line of research was economic; the use of superplastic metals enables certain components to be moulded quickly and efficiently, instead of needing to be laboriously machined.

Titanium alloys had been manipulated in this way for a long time, but what the scientists really wanted to know was whether they could produce superplastic steel. As a first step, they began to explore ultra-high-carbon steels; when mixed with steel, carbon reduces the size of the molecules that make up the metal. This is important, for superplastic metals have molecules up to 200 times smaller than those of ordinary metals.

When steel contains just over one per cent carbon, the result is a compound called cementite; this prevents the molecules from becoming too large. However, if the metal is brought to a high temperature, the iron next to the cementite begins to melt, and the metal shatters easily. This is what was frustrating the efforts of the mediaeval swordsmiths, although they did not know the technical details.

The swordsmiths' mistake had been to try and work the metal

while it was white-hot. What they should have done, according to the modern scientific findings, was to heat it to a temperature of 2000°F (1100°C), rolling it out continuously, and then cool it to 1200°F (650°C), keeping the metal at that temperature while it was being worked. The constant process of rolling ensures that the cementite compound has no time to settle down and become brittle, and the metal can then be worked at room temperature.

This was the process devised by the metallurgists, who suddenly realised that they had unwittingly hit upon the secret of Damascus steel. And the rediscovered technique was applied to a modern secret weapon: America's Rockwell B-1 supersonic bomber, which entered service in the latter years of the Cold War.

What is amaranth?

In their endless quest to find additional sources of food to bolster the world's overstretched crop resources, scientists have made an important discovery. Hundreds of years ago, crops of a nourishing, protein-rich plant called amaranth flourished in Central and South America. Cultivated with the methods in use today, amaranth could have provided half mankind with a life-giving staple diet of grain and cereal. Now agriculturalists are having to learn to cultivate it all over again, because four centuries ago European colonists deliberately destroyed it wherever they could find it in order to wipe out native populations.

Amaranth – the name is derived from a Greek word meaning 'everlasting' – flourishes in various forms practically all over the world; in fact, it covers one of the biggest plant families on Earth. Chickweed, a gourmet's delight for birds but a perpetual nuisance to gardeners, is part of the same species, and so is tumbleweed, which rolls across the screens of John Ford westerns.

The ancient Aztecs of Central America were the main cultivators of amaranth. The crops they grew stood shoulder-high, and their grain kept the Aztecs healthy. Apart from being high in protein, the plants contained an amino acid called lysine, which the Aztecs would not have known of, but which is essential to life, and absent from more traditional crops such as wheat and rice.

Tests carried out by the World Health Organization have shown that amaranth, mixed with other grain, almost completely fulfils the requirements for a healthy, balanced diet. In other words, it would sustain the human metabolism at a high level, even if nothing else were available to eat.

Why is it, then, that the potential of amaranth has been overlooked for so long? The blame seems to lie squarely on the shoulders of one man – Hernando Cortes, leader of the army of Spanish conquistadores that shattered the Aztec civilization for ever.

To the Aztecs, amaranth was more precious than gold, which to them had no significance other than as a pretty ornament. The last Aztec high king, Montezuma, used to demand an annual tribute of a quarter of a million bushels of amaranth from his subjects. The plant had great religious significance, too; mixed with honey, the grain was used in the central rites of various Aztec religious ceremonies.

None of these facts escaped Cortes when, in 1521, he began his programme of extermination against the Aztecs of Mexico. He soon realised that the quickest way to commit genocide was to destroy the Aztecs' source of food, and so, as well as burning native cities and murdering their inhabitants, he also burned their crops and turned their fields into deserts in which nothing would grow. The result was that amaranth virtually disappeared from the human diet. Its various forms continued to grow in the wild, but the secrets of successful cultivation died with the Aztecs.

It was not until the 1960s that scientists working in famine-stricken areas of Africa – mainly Tanzania – noticed that many of the healthier villagers were collecting and eating the seeds and leaves of a plant that grew profusely in the neighbourhood. It turned out to be a variety of amaranth. They made studies and reported their findings to government agricultural departments in the USA and Europe, but failed to awaken interest.

It has taken 30 years, and a serious worldwide shortage of essential foodstuffs, for experiments in growing amaranth to get under way. Cultivation is far from easy, and researchers are trying to compress all the development that should have taken place during the 'lost' 400 years into just a few growing seasons. There are a lot of things they might have learned from the Aztecs: how, for instance, did Montezuma's people grow amaranth more than 4 ft (1 m) high without the stems collapsing under the weight of the seeds? And how did they prevent the plant from shedding its seeds as soon as it matured, before they had time to harvest it?

There is no doubt that the cultivated amaranth of four centuries ago was a good deal tougher than strains cultivated so far under test. At present, scientists are trying to produce a hardy hybrid plant by crossing various strains. Some experiments have already seen promising results; a cross between types of amaranth from Africa and Mexico, for example, has produced a strain which is short in height, has a high grain yield, and is very resistant to insect pests and plant diseases.

The biggest obstacle in the way of turning amaranth into a crop of worldwide significance is that no-one yet knows how much it will cost to farm it on a commercial scale. Farmers are understandably cautious; 'wonder crops' of various kinds have been hailed by scientists many times before, only to turn out to be dismal commercial failures.

Possibly, amaranth's first application will be as a foodstuff for animals. When it has proven its value in this respect, agricultural experts might then decide that it is worth producing it to feed the two-thirds of the world who are less well fed than the average British cow.

What is a stroke?

Every year, on average, at least 100,000 people in Britain suffer what is commonly called a stroke. Of this total, 50,000 die. Only about half the remainder recover sufficiently to be able to look after themselves; 15 per cent are so disabled that they have to be confined permanently in a nursing home. In fact, stroke is the third most common cause of death in the UK, USA, and most developed countries, coming after heart disease and cancer. The incidence of new stroke cases is two per cent per 1000 of the population; there are about 500,000 a year in the USA, resulting in some 200,000 deaths.

Strokes present an enormous challenge to modern medicine, yet the outlook for a person suffering from a stroke is little better than it was 20 or 30 years ago. Brain scans conducted after a stroke can now tell the difference between the major types, but since very few types of stroke can be treated, this amounts to knowledge which, at the present time, is almost useless.

In very few cases of people who have survived a brain haemorrhage, skilled neurosurgery may help to reduce the chances of a stroke recurring, but in the vast majority of cases the best that modern high-technology medicine can do is to make the lives of victims a little more comfortable than was previously possible.

Strokes cannot be reversed; they can only be prevented. They are caused when a blood clot or haemorrhage cuts off oxygen from the cortex, or any other part of the brain, and kills the tissue. Sometimes a massive area of the cortex is affected, leading to death or total incapacitation, but if the oxygen is cut off from only a small area the result is a slight stroke in which the victim may, for example, lose sensation in an arm.

In that case, physical therapy over a period of months helps the patient recover some of the arm's functions. What happens is that different parts of the cortex control the nerve endings of specific parts of the body, so that if an area of the cortex is damaged by a stroke, the corresponding skin surface is numbed.

Eventually, the part of the cortex controlling an adjacent skin surface spreads into the lesioned area of the brain. New neurons are activated in the deadened surface, and if the stroke is slight, sensation is restored. However, the spreading action of the cortex is limited to only about 0.02 in (0.05 cm), so a large lesion results in less complete recovery. What scientists do not yet understand is how the brain adapts in this way; when they eventually unlock the secret of the brain's adaptation process, they may at last know how to proceed with the treatment for a stroke.

In the meantime, prevention is the only answer. The risk factors for a stroke are similar to those for heart disease, except that a person with high blood pressure is seven times more likely to have a stroke than someone with normal blood pressure. Keeping a close watch on one's weight is one of the main safeguards. Overweight is a principal factor in high blood pressure; even people who are 5 to 10 lb (2 to 4 kg) overweight may be at risk, and a person who is 20 per cent more than their desired weight may be considered obese and should receive medical treatment.

Scientists have found that over a period of 26 years, the death rate increases by two per cent for each pound over the desired weight. Being 10 lb (4 kg) overweight in your mid-forties means that you run a 20 per cent greater risk of suffering a stroke than someone whose weight is normal.

What is the sound barrier?

What we call sound is a vibration, a pressure variation in the air that communicates itself to the surrounding air and travels in every direction, spreading out as an expanding sphere. The changes in pressure detected by the human ear are very small; the variations in air pressure which accompany human speech are perhaps one-millionth of the normal pressure of the atmosphere, and a whisper, just detectable by a person with good hearing, would be made up of changes of pressure as small as one billionth part of one atmosphere.

It was Isaac Newton (who else) who first came up with a formula for the speed of sound. In his *Principia*, he stated that the velocity of propagation of a small pressure change is equal to the square root of the quotient of the pressure of the air and its density. Because Newton did not know that the compression of the air in a sound wave takes place so rapidly that the heat generated does not have time to get away – or, in technical language, that the process is adiabatic – his formula was a little bit out, putting the speed of sound about 15 per cent lower than it really is. The source of error in the calculation was pointed out later by the French mathematician Pierre Simon, Marquis de Laplace (1749-1827).

Newton's formula, subsequently modified, shows exactly how the speed of sound changes in the atmosphere. Since the temperature of the air varies with locality, altitude, season and time of day, the speed of sound is not nearly as constant as, say, the speed of light. The average value for the air temperature in the British Isles at sea level is 60°F (15°C), and from meteorological records we know with fair precision the average rate of fall of temperature with height. From this information, we can calculate that the speed of sound at sea level is approximately 760 mph (1220 km/h) at sea level, reducing to 735 mph (1180 km/h) at 10,000 ft (3000 m), 707 mph (1130 km/h) at 20,000 ft (6100 m), and 679 mph (1086 km/h) at 30,000 ft (9100 m).

The ratio of the speed of a body to the speed of sound is known as the Mach number, named after the Austrian physicist Ernst Mach (1838-1916). Mach One is reached when a body has a velocity equal to that of sound, Mach Two, a velocity twice that of sound, and so on.

When a body moves through the atmosphere at low speed, the air in its vicinity is able to take up the resulting changes of pressure without too much crushing together of the molecules, in other words, without noticeably increasing the local density. As the body approaches Mach One, however, the density changes in its vicinity are much more pronounced, because in this case the air meeting the body does not have sufficient time to conform to the new situation and is compressed into a gas of high density. As the speed of the body increases, it generates a pressure wave whose front is spherical, so that the whole disturbance grows like a soap bubble with the front advancing outwards towards the speed of sound. As the body itself approaches the speed of sound, its velocity generates a series of small pressure pulses, each of which imparts an acceleration to the air through which it passes. The result is that sooner or later a pulse catches up with the one that has gone before, so that eventually all the pulses pile up to form a sharp-fronted 'wave'.

In early attempts to achieve supersonic flight after the Second World War, aircraft designers found that these compression waves resulted in a large increase in drag at about Mach 0.9. Also, the shock waves generated at the nose of the aircraft tended to break back across the airframe, and some of these waves were travelling at supersonic speed even though the aircraft was still subsonic. As a consequence, aircraft approaching Mach One experienced severe buffeting, often accompanied by catastrophic failure of the airframe or, at best, loss of control. In the days before ejection seats were commonly fitted, this phenomenon cost the lives of a number of test pilots. The popular press latched on to the growing catalogue of disaster and talked of an impenetrable wall in the sky, a 'sound barrier' which nothing could break through. That, of course, was nonsense; bullets, shells and rockets like the German V-2 had all exceeded the speed of sound years earlier.

The real problem lay in the design of the aircraft, and the solution lay in fitting wings with an optimum sweepback of 35 degrees. If the

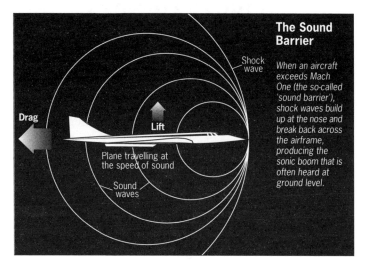

The Sound Barrier

Shock wave

Drag

Lift

Plane travelling at the speed of sound

Sound waves

When an aircraft exceeds Mach One (the so-called 'sound barrier'), shock waves build up at the nose and break back across the airframe, producing the sonic boom that is often heard at ground level.

wingtips of an aircraft are swept back, the shock waves generated at the nose of the aircraft miss them, so there is no complication of an interference between the wings and compressibility. Once this had been established, the road to supersonic flight was open, and the so-called 'sound barrier' had been broken. Later designs took aircraft to Mach Two and beyond.

And the 'sonic boom' that sounds like a thunderclap and occurs when an aircraft flies faster than sound? This is nothing more than the trailing edge of the cone-shaped shock wave reaching the ground. Sometimes, the 'pressure signature' of an aircraft in supersonic flight creates a shock wave from the nose and tail, in which case there is an audible 'double boom'.

What is rubber?

Rubber is indispensable to modern civilization. All kinds of applications, from tyres to rubber bands, depend on its resilience and elasticity, and in one way or another we all come into contact with it every day of our lives. Yet, like so many other essential things, we take it for granted, and seldom pause to consider what it is.

Natural (as distinct from synthetic) rubber is manufactured from the milk-like juice called latex that is present in the tissues of several species of plants. These grow mostly in tropical and sub-tropical countries, and the predominant species is *Hevea brasiliensis*, a wild variety found in the rainforests of the Amazon region of Brazil. The rubber produced from it is known as Para rubber. Latex is still tapped in the wild in this area because efforts to establish plantations have largely failed. Trees transplanted to Indonesia, Sri Lanka and Malaysia produce most of the modern world's natural rubber.

The tree itself grows to a height of about 60 ft (20 m), with a trunk of about 6 ft (2 m) in girth. The leaves are rather like those of a horse chestnut, but with only three leaflets radiating from the top of the stalk. The flowers are small, green and sweetly scented and are arranged in panicles, like oat ears, with the female flowers at the top and the males below. This arrangement helps to bring about cross-pollination, producing more and better seed. The fruits are three-lobed, with a single seed in each lobe; the seeds are rich in oil very similar to linseed oil.

The latex is found in the inner bark of the tree, and is tapped by cutting or shaving the bark with a very sharp knife. Tapping is usually done in the morning, when the flow of latex is most abundant. It is then coagulated, and tappers in Brazil use a very old technique that involves smoking the latex over a fire. The thickened latex is then poured over a stick or paddle on which it coagulates to form a ball or cake. Latex gathered on modern plantations is coagulated by the addition of chemicals such as acetic acid; these cause the rubber to form a sheet on the surface of the liquid, which is then pressed between rollers to remove excess moisture. An alternative process involves spraying the crude latex into a stream of hot air which evaporates the liquid, leaving particles of rubber behind.

Rubber was first described scientifically by two Frenchmen, C-M de la Condamine and François Fresneau, following an expedition to Latin America in 1736. The substance was given the name rubber in 1770 by the English chemist Joseph Priestly when he found it could be used to rub out pencil marks.

In the early years of its commercial use, rubber was plagued by a tendency to soften in hot weather, to become brittle when it was cold, and to turn sticky when it was exposed to solvents. The problem was not solved until, after years of experimentation, the American inventor Charles Goodyear discovered a technique for hardening rubber and making it more elastic by combining it with sulphur and white lead and exposing it to heat. He discovered the process, which he called vulcanization, in 1839 and patented it in 1844. It was improved in the early 1900s by the introduction of organic chemical accelerators and antioxidants, which helped the rubber to cure more quickly and resist a tendency to become brittle with ageing. Later, strength and durability were increased by the introduction of reinforcing agents such as carbon black.

There is a remarkable story behind the introduction of rubber to Sri Lanka and Malaysia. It began in 1875, when a botanist called Henry A. Wickham delivered 70,000 Para rubber seeds which he had collected in Brazil to the Botanic Gardens at Kew, London. The seeds were quickly sown in plant houses and 3000 germinated. Two months later, 1900 of these were shipped to Sri Lanka (then Ceylon), and it was from this small beginning that the vast rubber industry of the region was built up.

Natural rubber occurs in various parts of the world, but none of it can compete with the Para variety. In Africa, most of the rubber-yielding plants belong to the Periwinkle family, *Apocynaceae*. The best is the Funtumia rubber tree, *Funtumia elastica*, and several species of *Landolphia*. Rubber is also produced from a shrub called Guayule, *Parthenium argentatum*, and some species of dandelion, especially one named *Taraxacum kok-saghyz*, from southern Russia.

The production of natural rubber continues to hold its own in the world, despite competition from synthetic rubbers. The production of the latter was greatly stimulated by the Second World War, when the Far East's natural rubber resources were mostly captured by the Japanese. Synthetic rubbers are based on polymerization processes, which are chemical unions of two or more molecules of the same kind to form a new compound.

What is soya?

The soya bean – or soybean, to give it what is probably a more correct name – is one of the world's oldest cultivated foodstuffs. It is also one of the most important, a fact that has been recognized in the east for many centuries though in the western world only comparatively recently.

The soybean is a native of China, and according to tradition its cultivation was begun about 5000 years ago by the Emperor Shennung, the father of Chinese agriculture. It was introduced to Europe by French missionaries in 1740 and grown successfully in the Paris botanic garden. In 1804 it arrived in the USA with early Chinese immigrants, and became an important crop in the South and Midwest in the middle of the 20th century.

Botanists believe that the soybean plant derived from *Glycine ussuriensis*, a legume native to central China. The plant looks like a rather hairy dwarf French bean with tiny flowers and short, hairy pods containing three or four beans. An erect, branching plant, it varies in height from just a few inches to more than 6.5 ft (2 m). Its flowers, which are self-fertilizing, are white or a shade of purple, and the seeds can be yellow, green, brown, black or bicoloured.

The soybean is the most nutritious and easily digested food of the entire bean family, and is one of the richest and cheapest sources of protein, forming a leading staple in the diet of people and animals across the world. The pale kinds of bean are the most suitable for human consumption and are unique among vegetables in that they provide almost a complete diet in their own right. The seed contains 17 per cent oil and 63 per cent meal, 50 per cent of which is protein. The bean contains no starch, which makes it exceptionally valuable as a food for diabetics.

The Chinese have devised many ways of preparing the beans in addition to simply cooking them in water. After boiling, the beans may be ground up with water to produce a milk every bit as nutritious as cow's milk; factory-produced soybean milk is widely consumed in China, where it has been available for years in bottles and cartons, and it can now be readily purchased in the west. Soy milk is curdled by the addition of salts or rennet, just as animal milks are, and the curd which separates is pressed and made into *tofu*, which looks rather like cottage cheese.

To the average westerner, the most familiar soybean product is soy sauce, which is made by fermenting soybeans and rice with a liberal quantity of salt for anything from eight months to five years. Soybeans are also sprouted for use as a salad ingredient, and may be eaten roasted as a snack. Sprouting soybeans develop vitamin C, which is normally lacking in them.

The soybean will grow in most types of soil, but it thrives in warm, fertile, well-drained sandy loam. The plant matures in September and October, when harvesting takes place.

The part played by the soybean in today's world continues to expand as new applications are found for it. The oil is used in salads and confectionery, as a high-protein meat substitute in many products, including baby foods, and can also be processed into margarine, shortening, and vegetarian cheeses. In industry, it is used as an ingredient in the manufacture of paints, varnishes, adhesives, fertilizers, insect sprays and fire extinguisher fluids, among many other products. The USA is now the world's leading producer, followed by Brazil and the People's Republic of China. Emperor Shen-nung certainly did the world a great service.

What is hypothermia?

Years ago, in the days before adequate heating systems, people wrapped up well before they went to bed in the winter. They might not have looked alluring, but few of them died from hypothermia in their sleep. Today, sadly, the opposite is often the case. Every winter, an alarming number of people – mainly the elderly and undernourished – die miserably in their own homes from this condition.

The tragedy is that many of the victims do not realize what is happening to them, and consequently fail to take proper precautions in time to save themselves. In fact there is nothing mysterious about hypothermia, and potential victims can safeguard themselves against it by following a few simple guidelines.

Basically, hypothermia is a state of low body temperature. Normally, the human body maintains a fine balance between the amount of heat it produces, either by physical activity or the intake of food, and the amount it loses through the evaporation of perspiration. If the heat loss exceeds the heat gain, the body temperature falls. This in itself is not serious, because the body has certain built-in mechanisms which are automatically brought into play to compensate for it. The act of shivering, for example – over which we have no control – produces muscular activity, which in turn produces more heat. Also, when the body is exposed to cold, the small blood vessels that supply the skin automatically contract through a reflex action of the nervous system.

This is a very powerful safeguard, because it diverts the blood circulation away from the skin and the cold areas and effectively turns the skin into a self-sealing protective layer that traps the body's heat inside. When this happens the skin's heat conductivity drops to about the same level as that of cork, which is very low. In fact, cork was once used to insulate vacuum flasks for this very reason.

Sometimes, however, the body is unable to cope with the heat loss, resulting in what is known as accidental hypothermia. There are two types, one of which occurs out of doors and strikes healthy individuals, for instance climbers who are exposed to severe cold for prolonged periods and who are inadequately insulated; or people who are accidentally immersed in cold water.

The other type is urban hypothermia, which occurs indoors and mainly strikes the very young or the very old. The young are at risk because their nervous systems are not yet properly developed, while the elderly suffer because their nervous systems are degenerating. In some cases, alcohol, drugs or illness can also interfere with the mechanism that regulates the body's temperature.

Hypothermia starts to take effect when the body temperature begins to fall below the normal level of 98.6°F (37°C). The symptoms are excessive shivering and low vitality. Below 95°F (35°C) the flow of blood to the brain decreases and the victim has difficulty in talking and walking. Below 86°F (30°C) the nervous system begins to break down, protective shivering ceases, and the muscles become rigid. Eventually the victim becomes comatose, and when the body temperature falls below 77°F (25°C) breathing may cease. At any point below 35°C the victim needs emergency treatment. Some simple precautions can prevent this all from happening in the first place; adequate clothing is the obvious answer.

Many elderly people wrap themselves up well indoors, but neglect to cover their heads – yet half the heat loss from a fully clothed adult is through the head. The feet are another major heat loss area. Our ancestors of a century or more ago realized this, which is why they wore nightcaps and bedsocks in addition to voluminous nightshirts.

In the Middle Ages, people who lived in cold, stone houses subject to severe winter weather – on the Scottish borders, for example – knew the secret. They 'bundled' – in other words, they lived together in a communal room around a fire that was kept going day and night, and stayed there until the worst was over.

They ate well, too, though their food was not extravagant – wheaten bread, mutton stew, pork and of course porridge. But it was the kind of food that kept their energy levels high, and mostly they survived well on it. The fat content of their daily intake would make today's dieticians shudder – but it is a fact that people with a substantial layer of body fat lose heat more slowly, and are always less susceptible to extremes of cold.

What is a 'sixth sense' – and do we all have one?

A sk anyone how many senses a human being possesses, and nine times out of ten the answer will be five: vision, hearing, touch, smell and taste. As well as these, however, we have a sixth sense called proprioception or kinaesthesia, which subconsciously keeps us informed of the relative positions of our limbs, the tensions in our muscles and so on.

So what we have always thought of as a 'sixth sense' – a kind of hunch that tells us that something is about to happen – is really nothing of the sort. It's a seventh sense.

There is no doubt that such a thing exists; animals display it frequently, and humans certainly have it too, although they have forgotten how to exploit it. Our remote ancestors, who lived by their wits and whose senses were finely tuned to the natural world, must have had an instinctive feeling for anything that was out of place in their environment, and a premonition of any coming change that was likely to affect them.

If proof of such a phenomenon is needed, all we need to do is observe the animal world. Animals, for example, often seem to know when their owners are coming home; there are many recorded instances of free-roaming dogs and cats returning home in time to greet their owners, even though the homecoming might not be at a specific time of day.

Some researchers into the subject believe that this seemingly paranormal attribute is a product of the distant past. For millions of years, among the wild ancestors of dogs and cats, the young remained behind while the adults went off hunting, and the return of the hunters with food was an event of the most vital importance, accompanied by a complex greeting ritual which continues to be displayed by domestic pets today. After all, both species have a close relationship with humans that goes back at least 10,000 years.

But if the sixth (or seventh) sense of animals is a product of long evolutionary habit, rather than some form of extra-sensory perception, what about the alleged ability of animals to sense oncoming earthquakes? One possibility is that animals are sensitive to vibrations of the earth so minute that our instruments fail to detect them. Geologists now know that months, even years before a major earthquake develops, rocks start ringing like an alarm bell; they call it the coda sequence. The rocks begin to vibrate when the first tremors affect them, and this vibration is easily detected by simple seismic equipment. The duration of the ringing, the coda, depends on the extent to which the rocks deep underground are cracked and dislocated. It may be that animals can sense these, and more subtle vibrations.

Another possibility is that animals respond to the dramatic increase in static electricity which precedes an earthquake. Humans can sense this too, and in some it causes a vague ache or throbbing in the head. A third explanation is that animals respond to sudden shifts in the Earth's magnetic field; perhaps they respond to all three phenomena at the same time. What is certain is that animals, particularly dogs and cats, become intensely agitated just before an earthquake strikes, and often flee in panic into the open. Cats with kittens sometimes carry them out of doors to a place of safety. Birds also stop singing and take to the air.

A similar 'extra' sense is almost certainly latent in humans. Some people have the ability to sense the direction of magnetic north, which means that they must be attuned in some way to the Earth's magnetic field; others know exactly what time of day it is without reference to a timepiece. And we have all had the experience, from time to time, of sensing that we are being looked at.

What is a hurricane?

A hurricane is a very severe form of cyclone, or depression, which is an area of low atmospheric pressure. In the Pacific, hurricanes are known as typhoons. On average there are about 80 hurricanes every year; they cause some 20,000 fatalities and are responsible for an enormous amount of damage to property. They also present a serious shipping hazard, due to the combined effects of high winds, high seas, flooding from heavy rainfall and coastal storm surges.

A typical hurricane system has a diameter of about 400 miles (650 km), although Pacific typhoons can be very much larger. The central pressure at sea level is usually 950 millibars (mb), dropping occasionally to less than 900 mb, while sustained surface winds can be as high as 200 mph (320 km/h) although more commonly they vary between 74 mph (118 km/h) and 120 mph (190 km/h).

Hurricanes develop in the western sections of the Atlantic, Pacific, and Indian Oceans between 5° and 20° north or south of the equator. The main activity in both hemispheres is in late summer and autumn, but smaller cyclonic storms can affect both the western North Atlantic and North Pacific areas as early as May and as late as December. In the Bay of Bengal, where hurricanes have caused appalling loss of life, there is also a period of activity in early summer.

Hurricanes rarely form near the equator, where the Coriolis effect (the effect of the Earth's rotation on the atmosphere, which deflects wind currents to the right and left in the northern and southern hemispheres respectively) is at a minimum, or beneath jet streams, where there is a powerful vertical wind shear. Exactly how and why they form is still being studied, though it is known that a number of conditions are necessary for their formation, one of which is a large area of

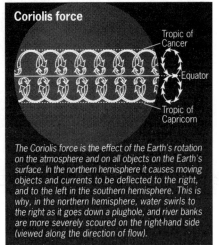

The Coriolis force is the effect of the Earth's rotation on the atmosphere and on all objects on the Earth's surface. In the northern hemisphere it causes moving objects and currents to be deflected to the right, and to the left in the southern hemisphere. This is why, in the northern hemisphere, water swirls to the right as it goes down a plughole, and river banks are more severely scoured on the right-hand side (viewed along the direction of flow).

ocean with a surface temperature of more than 80°F (27°C). There is also a connection, as yet unexplained, between hurricane formation and the equatorial trough, an area of low pressure that changes its position seasonally.

Hurricanes are thought to be born out of the convection (heat energy transferring) cells in a build-up of cumulus clouds, where a release of energy organizes the cloud into spiral bands that flow outwards. This flow in turn leads to the development of very low pressure and high wind speeds near the surface. One notable feature of a hurricane system is its warm vortex, since other tropical depressions and incipient storms have a cold, showery area at their centres. The central, calm area of a hurricane system is called the eye; it has a diameter of 20-30 miles (30-50 km), and the winds spiral anticlockwise around it. The warm core develops through the action of up to 200 cumulonimbus cloud towers releasing latent heat of condensation. The tops of the massive cumulonimbus clouds that accompany the hurricane may rise to 40,000 ft (12,000 m).

A hurricane, then, forms as the result of an atmospheric disturbance which, if the conditions are favourable, develops into a tropical depression and then a tropical storm, lasting up to five days. Its main source of energy is latent heat derived from condensed water vapour, which is why hurricanes form and gather strength only over areas of warm ocean. The tropical storm is transformed into a hurricane by the release of latent heat energy in the cumulonimbus clouds, which sets up an anti-cyclonic cell in the upper troposphere. Winds flow outwards from the centre of the system at high level and return at low level, so that potential energy is constantly being generated and transformed into kinetic energy. A hurricane system travels at a rate of 10-15 mph (16-24 km/h), the speed being dictated by the rate of movement of the upper part of the warm core.

What is salmonella?

The question should really be what *are* salmonella, since there are about 1500 known species in this genus. At any rate, salmonella has caused a good deal of paranoia in recent years, largely because the upsurge in foreign travel has brought with it an attendant risk of food poisoning. Yet scientists know all about the group of bacteria that causes salmonella poisoning, and how to prevent it. It therefore seems surprising that outbreaks of food poisoning caused by salmonella continue to increase year after year.

Salmonella bacteria can be divided into two broad groups. One of these causes typhoid and paratyphoid fevers, while the other is responsible for salmonella food poisoning. Salmonella enters the food chain mainly from animal sources. It starts life in a field with a host of other microbes – most of which we have learned to live with – and the infection passes to the animal through its fodder. Breaking down this cycle of infection would be a long and costly business, requiring radical changes in farming methods.

Where the cycle can be broken is at the point where salmonella passes from animal products to human beings, and scientists agree that the main problem here is a lack of public education in how to treat suspect foods. Food catering hygiene, or lack of it, is a principal cause of infection. Although there have been outbreaks in restaurants, hotels and hospitals, the greatest risk of all lies in the home, where hygiene standards can often be very poor. The risk has increased with a growth of enthusiasm for home-cooked Oriental and Middle Eastern foods. In April 1988, for example, about 100 people in the south of England suffered salmonella poisoning caused by infected bean sprouts. These are cultivated in warm water, which is an ideal breeding ground for the bacterium, but the flash-frying method normally used to cook the sprouts was not, in this instance, effective in killing the bacteria. Ideally, the sprouts should have been boiled for at least 15 seconds before being fried.

Similarly, salmonella bacteria in eggs can be killed by thorough cooking. In milk, pasteurization is the answer. This is a process, named after Louis Pasteur, in which milk is heated to 161°F (72°C) for 15 seconds, and then rapidly cooled to 50°F (10°C), killing all bacteria or delaying their development. Almost all milk in Scotland has been required to undergo pasteurization since 1983. In the three years before that, there were 14 outbreaks of salmonella poisoning caused by infected milk; in the years since, there have been none.

Common household faults that can lead to salmonella poisoning include the preparation of food too far in advance, undercooking, and poor storage. Prepared meals should always be refrigerated at between 35°F (2°C) and 41°F (5°C), and simmered for at least five minutes during reheating, with constant stirring to eliminate any cold spots. Perhaps most important of all, raw poultry should be kept separate from other foods.

Salmonella will continue to be a nuisance, and a potential health hazard, until scientists come up with a fast and foolproof way of detecting the bacterium in the early stages of food processing. Various processes already exist to detect and eliminate it effectively, but they are lengthy. In the meantime, the answer to salmonella control lies literally in our own hands.

What is radon?

Forget the worries about nuclear power. Thousands of people all over Europe and the USA are breathing air with radiation levels that would not be tolerated inside a nuclear power station. The threat comes from a radioactive gas called radon, which bubbles up out of the ground. According to some estimates, radon is the biggest single cause of lung cancer after smoking, killing up to 1000 people in Britain every year.

It is only fairly recently that the world has woken up to the perils of radon. The Americans were the first to sound the alert in December 1984, after a man named Stanley Watras set off radiation alarm bells on his way *into* work at a nuclear power station near Philadelphia. He continued to set off the alarms every morning for a fortnight, until investigators discovered the source of his radioactivity. It was Watras's own home, which was built on top of an old vein of uranium 30 ft (9 m) wide. The decaying uranium deposits were releasing radon gas, which was leaking into the house. Later, one of the investigators said that breathing the air in Watras's house was as risky as smoking 135 packs of cigarettes a day. He had to leave his home for six months while the authorities found ways of making it safe.

Many people are exposed to some form of natural radiation, the main source being cosmic radiation, which filters to Earth from outer space, and the Earth's rocks. In Britain, the most important underground source is radioactive isotopes, decaying from uranium on the fringes of granite areas such as Somerset, the Derbyshire Peak District, Northamptonshire, Clwyd, West Yorkshire, Shropshire, Gloucestershire and Lincolnshire.

Radon, which is colourless and odourless, is the densest gas known, and occurs naturally when a radioactive isotope, radium 226, disintegrates underground. The gas then bubbles up out of the ground, making its radioactivity dangerous to anyone breathing it. Its half-life – in other words, the time it takes for the radioactivity to decay to half its original value – is nearly four days. The amount of radon that escapes from rocks and reaches surface soil depends on a number of things. It has a head start if the rocks are close to the surface, and if fissures or faults are present. Old mines serve the same purpose, which is why Devon and Cornwall are the worst affected areas in Britain, with radon concentrations 15 times higher than the national average.

Radon is measured in Bequerels per cubic metre. The national average is 22 Bequerels, but in some places, especially in the West Country, the reading is 1000 Bequerels per cubic metre or higher. Health authorities in Britain and elsewhere generally accept that homes where the radon level exceeds 400 Bequerels per cubic metre are at risk, but even this is four times the radiation safety level set for workers in most nuclear power stations.

Tracking down high concentrations of radon is no real problem; the chemistry of the gas has been well researched over the years by mineral firms prospecting for uranium and other ores. The presence of radon on the surface is a good indication that uranium is present deep underground. The British Geological Survey has maps, both published and unpublished, which have been used to pinpoint the communities thought to be most at risk.

Coping with the radon threat is not easy. Researchers tried fitting one badly affected house with a leak-proof floor, only to find that the gas was seeping up through a porous internal wall. One step that can be taken is to stop building houses where there are heavy radon concentrations, but here too there are problems. Researchers were puzzled to find radon present in some new houses a long way from badly affected areas in Cornwall, until they discovered that the building firm had used hardcore brought from spoil tips from a derelict tin mine many miles away.

What is St Elmo's fire?

July 11, 1881. During the middle watch the Flying Dutchman *crossed our bows. She first appeared as a strange light, as of a ship all aglow, in the midst of which light the masts, spars and sails, seemingly those of a normal brig, some two hundred yards distant from us, stood out in strong relief as she came up. The look-out man on the forecastle reported her as close on the port bow, where also the officer of the watch from the bridge clearly saw her, as did also the quarter-deck midshipman, who was sent forward at once to the forecastle to report back. But on arriving there, no vestige nor any sign of any material ship was to be seen...*

So reads the log of the Royal Navy warship HMS *Bacchante*, cruising in the Pacific. It records one of the many sightings of phantom ships that have occurred over the centuries; sightings of spectral vessels whose masts and rigging were surrounded by a halo of blue fire. The *Flying Dutchman* is the most celebrated of them all, and the legend of the ghost ship is said to have its origin in one of two real-life characters. The first was a Dutch sea captain called Vanderdecken, a contemporary of the 17th-century Admiral Tromp, who was condemned to sail the oceans forever because he was a notorious blasphemer; the second was a mariner called Bernard Fokke, who allegedly made a pact with the devil so that his ship could reach the East Indies in a record time of 90 days.

All colourful stuff, forming an indispensable part of the lore and superstition of the sea. But what rational explanation might lie between appearances of the *Flying Dutchman* and other maritime phantoms?

In all probability, the sightings have involved nothing more ghostly than solid, honest-to-goodness merchant ships with their masts and spars glowing with a phenomenon called St Elmo's fire, a bluish, flamelike electrical charge that sometimes builds up around ships' masts or other pointed objects such as radio aerials – and even human fingers – in stormy conditions. It also manifests itself around aircraft. It gives off a steady glow, and although it is high voltage, it is low current, which makes it harmless. St Elmo, incidentally, sometimes called St Erasmus, is the patron saint of sailors; he was an Italian bishop who was subjected to a rather nasty martyrdom in the fifth century.

Terrestrial electrical charges of this kind also occur around power lines, ropes, or even falling snow. Electricity leaks into the air from mountain tops, producing streamers of light or a steady yellowish or greenish glow; this is often experienced in South America, and is known as the Andes glow. Streamers of glowing light reach high into the sky and are frequently visible for hundreds of miles around.

St Elmo's fire is easily explained, but it sometimes produces strange and alarming effects. One American meteorologist, carrying out observations in a heavy snowstorm, told how his fingers became tipped with blue cones of 'cold fire' 3 in (8 cm) long; they were completely painless and gave off a whispering, or rushing noise. His coat cuffs also developed circles of blue light around them, and a blue flame enveloped his moustache. The effect vanished when the snow stopped falling, accompanied by a flash of lightning and a thunderclap.

What is electricity?

The term 'electricity' encompasses all phenomena caused by an electric charge, whether static or in motion. The elementary particles that are the basic units of electrically charged matter are the electron, which is negatively charged, and the proton, which carries a positive charge to an equal but opposite extent. Matter which contains an equal number of protons and electrons is electrically neutral; matter which contains an excess of electrons possesses an overall negative electrical charge, and similarly, matter which has an excess of protons possesses an overall positive charge. Two bodies both with positive or both with negative charges repel each other, while bodies with opposite or 'unlike' charges attract one another. The region in which these forces act is called an electric field.

An electric charge can be generated by friction or chemical change, and manifests itself in an accumulation of electrons (negative charge) or a loss of electrons (positive charge) on an atom or body. Atoms themselves do not hold a charge, but under certain circumstances they can gain electrons to become negative ions or lose them to become positive ions. This takes place during chemical reactions, or as a result of exposure to some forms of radiation. What we call static electricity – the phenomenon that makes sparks fly when we change nylon or silk clothing or brush our hair, or through mechanical force such as rubbing – is caused by a gain or loss of electrons from the surface atoms.

Electric potential is the work that is necessary to bring a positive electric charge over a distance to a certain point. If two points have a different potential there is said to be a potential difference between them, and if the points are joined by an electric conductor an electric current will flow between them. Current is measured in amperes, and flows through a conductor when there is an overall movement of electrons through it. The current at any point in the circuit is the amount of charge flowing through it. A conductor is any substance which allows the passage of electricity, such as a metal like copper, which is an excellent conductor; substances which do not, such as rubber, are called insulators. Current carries electrical energy from a power source – for instance a battery – to the components of the circuit, where it is converted into heat, light or motion. Direct current, (DC), such as that produced by a battery, flows in one direction only; alternating current (AC) flows in one direction until it reaches a maximum, decreases, then flows in reverse until it reaches a maximum in this direction too, at which point the cycle is repeated. The number of cycles completed per second is called the frequency.

The advantage is that its voltage (degree of electric potential) can be raised or lowered by a transformer; the efficient generation of AC at a power station requires high voltage, as does transmission via power lines, but low voltage is essential for safe usage. Power stations generate electricity at about 25,000 volts, but transmission requires 400,000 volts or more.

The name 'electricity' was coined by William Gilbert, physician to Queen Elizabeth I. It is derived from the Greek word for amber, a substance that attracts light objects when it is rubbed. The Romans are said to have known about this property of amber, and it is possible that the practical use of electricity goes back a very long way in history. There is a very interesting story that supports this supposition.

In 1936, archaeologists unearthed a strange object from the ruins of a village near Baghdad. It was a clay jar, containing a cylinder of sheet copper with an iron rod suspended in its centre. Similar objects were later found at other sites in Iraq. The original jar and its contents were put on display in the Cairo Museum. On numerous occasions, people who saw it remarked that it looked just like an electric cell battery, but no-one tried to establish this until 1976, when German scientists at Hildesheim built an exact replica and, as an acid substitute, filled it with grape juice. There was no longer any doubt that it was a primitive battery, for the chemical reaction that took place produced an electric current of up to two volts in strength.

Yet the original battery was believed to be as much as 2000 years old. So where did the people who lived near what is now Baghdad obtain the knowledge to build it – and for what purpose was it used?

The first question is impossible to answer, but the German scientists thought they might have the answer to the second. In an experiment, they immersed a small silver statue in a gold cyanide solution and passed an electric current from their replica battery through it. In just a couple of hours, the process had given the statue a thin layer of gold. It is possible, then, that ancient goldsmiths used electric current to electro-plate their valuables.

What is 'red sky at night'?

There is nothing quite as lovely as a sunset – especially in the winter, when the Sun is an orange ball in a frosty sky and the ground snow-covered. But even sunsets have changed a lot in recent history. If we could go back in time just a hundred years or so, we would see a completely different spectacle. The setting Sun would be bright yellow, and the sky around it varying shades of orange. As the Sun sank below the horizon, the colours adjacent to it would slowly change from orange to pale green and then deep blue, with low-lying clouds continuing to reflect the Sun's light for some time after it had vanished.

You can still see sunsets like that, but you have to go a long way to find them – to the few spots left on Earth which are relatively free of pollution.

That's the trouble with today's sunsets. They may appear magnificent, but in fact they reflect the ever-growing levels of pollution in the Earth's atmosphere. In industrial areas, where the atmosphere is laden with pollutants in the form of aerosols, the setting Sun is orange-red and the surrounding sky shows different shades of red, the bands of colour revealing the density of atmospheric pollution. If the level of pollution is very high, the Sun shows through as a dull-red disc that may fade away before it touches the horizon.

Admittedly, not all spectacular blood-red sunsets are caused by man-made pollution. On 28 Mar 1982, El Chichon – a volcano in a remote part of southern Mexico – suddenly exploded, hurling enough debris into the atmosphere to have an effect on the global climate for years. Sunsets around the world were literally set ablaze by volcanic gases blasted 16 miles (25 km) up into the stratosphere. Many other volcanic eruptions in recorded history have produced similar effects.

One of the most devastating occurred on 27 Aug 1883, when the volcanic island of Krakatoa, in the Sunda Strait between Java and Sumatra, exploded with a power equivalent to 1000 million tonnes of high explosive. For months afterwards, people all around the globe were treated to the sight of strange blue and red sunsets, caused by fine particles of drifting volcanic ash.

Industrial haze contains similar particles of dust and aerosols. Over a period of time, these settle into layers, the larger particles forming the denser layers closest to the ground. As the Sun sinks through these layers, its light gradually fades and takes on an orange-red colour.

This is because the particles 'scatter' the Sun's light. In a clear, unpolluted atmosphere, the scattering is caused by oxygen and nitrogen molecules, which are smaller than the wavelength of visible light. This is made up of the colours of the spectrum, as we see them in the rainbow. When the short-wavelength colours of the spectrum – violet, indigo, blue and green – are filtered out by particles, the Sun's light appears yellow.

Sunlight that passes through many layers of industrial haze loses still more of its intensity. The yellow light is also scattered, leaving the long-wavelength red-orange part of the spectrum visible to the naked eye.

Anyone can gain a good idea of how much pollution is present in the air by observing the sunset. Sometimes, just before the Sun sinks to meet the horizon, you can see a bright red glow surrounding the duller red of the Sun's disk. This glow, called the aureole, is caused by large dust particles floating in a dense layer close to the Earth's surface. If there is a shower of rain before sunset, the glow disappears because the moisture has brought the particles down to the ground.

The sunrise is also a good indicator of the level of pollution present in the atmosphere. Scattered light is visible before the Sun appears over the horizon, and its colour is dictated by atmospheric conditions. Vivid yellow, purple or deep blue in the sky before sunrise indicate that the atmosphere in the east is comparatively free of pollution, and when the Sun rises most of the sky becomes blue, with only a narrow orange-yellow band close to the horizon.

Atmospheric pollution has made the old saying 'red sky at night, shepherd's delight' pretty meaningless today. It belongs to the days before industrial pollution, when red skies at sunset were caused by dust particles rising from the land on currents of warm air, a sure sign of continuing good weather. A red sky at morning, on the other hand, was created by water droplets scattering the light of the rising Sun, indicating bad weather on the way.

Nowadays, red skies are just another sign that we are slowly choking the world to death.

What is the missing link?

Until about the middle of the 19th century, scientists assumed that man had always been man, a belief fuelled by the supernatural story of creation and the biblical allegory of Adam and Eve. Then, in 1857, an extraordinary fossil was discovered in the Neander Valley near Düsseldorf, Germany. Neanderthal man threw a spanner in the works of scientific knowledge, because although the fossilized bones were clearly different from those of *homo sapiens* – modern man – they were equally different from an ape's. A couple of years later, Charles Darwin published his theories of evolution in *The Origin of the Species* – and the search for mankind's remote ancestor was on.

For a long time, it was thought that the common ancestor of ape and man lived 20 million years ago, but in the 1970s and 1980s genetic studies of humans and chimpanzees, which share nearly 90 per cent of common genes, showed that the species separated much later – about six million years ago.

So what was the common ancestor? There are two likely candidates, both of which fit the time scale. The first is *Graecopithecus* (Greek Ape) and the second is *Dryopithecus* (Oak Ape), which are known to have lived in Greece and continental Europe about nine million years ago.

Early on, scientists thought that the missing link might have been a little ape-like creature called *Ramapithecus*, dating from around 14 million years ago. Its fossils have been found in East Africa and Pakistan. In the days when the common ancestor was thought to have existed, around 20 million years ago, Ramapithecus dropped into the gap quite nicely. But later, a close examination of the creature's facial structure showed that it resembled that of an orang-utan more than it did man's, and man and the orang-utan are on completely divergent tracks.

It was in the early 1990s that anthropologists announced a discovery which may be the key to the mystery. The find involved the fossil fragments of a creature that died four and a half million years ago near what is now the village of Aramis, in Ethiopia. Over a two-year period, the fragments were painstakingly assembled by a scientific team led by Professor Tim White, of the University of California. The scientists believe that the creature, which they named *Australopithecus ramidus*, is mankind's remote ancestor, the 'missing link' between men and apes. It has features in common with both.

The fossil is at least a million years older than that of any other candidate for the post of humanity's ancestor. The fragments consist of teeth, pieces of skull, arm bones, and also a partial lower jaw, thought to have belonged to a child. The adult creature was probably no more than 4 ft (1.2 m) tall, and it is not known whether it walked upright or went on all fours.

The previous oldest known human-like fossil, *Australopithecus afarensis* – known also by the much simpler name of Lucy – was found in 1975 at Hadar, about 50 miles north of the Aramis site. Scientists, comparing teeth from both fossils, established that those of the later find have thinner enamel and a different shape, suggesting a diet of leaves and fruit.

At first sight, *ramidus* might be mistaken for a chimpanzee, but it is not one. A careful examination of the cranium, teeth and arm bone fragments have shown that when this creature walked the earth, apes and the species that was to become man had already diverged from their common ancestor and were following separate paths of evolution.

No-one yet knows what led to the evolution of the two different species from the common ancestor, and there are other riddles also to be solved. About 1.8 million years ago, the creature that became man was distinctly ape-like, and yet within 200,000 years – a mere fraction of time in evolutionary terms – it had developed into a human being. Scientists are still trying to work out what caused this evolutionary spurt.

Another riddle is why apes hate water, and yet humans are able to swim from infancy. Some scientists are convinced that at some time in our prehistoric past, we went through a marine period when our ancestors lived mostly in water. Unlike apes, we are equipped with diving reflexes and we eat fish; we have flexible spines, which go with adaptability to water – but no ape can bend backwards.

Apes do not shed tears, either. Man is the only primate who floods his eyes with salty liquid to express certain emotions – perhaps another sign of an aquatic past.

What is rust?

The ancient Greeks had a technique for most things, and one technique involved dealing with the ravages of rust. Although their masons were superb craftsmen, their buildings needed to be extra strong to withstand the earth tremors that were not infrequent in the Balkans, and so they used wrought-iron cramps to join together blocks of stone and marble. They were clever, though, in that they encased the iron cramps in lead, so that they were shielded from the atmospheric conditions that cause rusting. The Acropolis, which has been around for the best part of two and a half thousand years, is visible proof that the method worked.

Rust, the reddish-brown oxide of iron formed by the action of moisture and oxygen on the metal, is something that we have to live with. We can take steps to prevent it, or at least to stave off its harmful effects for as long as possible, but as every car owner knows, it catches up with us sooner or later. First of all a few spots appear, accompanied by discolouring and staining of the paintwork; later on, parts of the vehicle become dangerously weak as corrosion sets in.

Corrosion is the erosion and eventual destruction of iron and steel by the chemical attack of the hydrated iron oxide that forms rust. The rate of corrosion is increased when the atmosphere is polluted by sulphur dioxide, the pungent gas produced by the burning of sulphur in the atmosphere; it occurs in certain industrial processes and is a major cause of acid rain. The corrosion process is also accelerated by salty air and road conditions, which is why vehicles should always be thoroughly hosed down, especially underneath, after travelling over winter roads where salt has been sprinkled to remove or prevent ice. Corrosion is mainly an electrochemical process, and acidic or salty conditions lead to the establishment of electrolytic cells on the metal, which cause it to be eaten away.

The gradual loss of parts of a car or other metal structure through rusting is called wastage corrosion. While the process is under way, the rust gradually occupies more space than was occupied by the metal that is being consumed; this increase in volume is due to the fact that the rust consists not only of the original metal atoms, but also of atoms from the atmospheric agents that sustain the process. When this happens, corrosion releases enough energy to deform and fracture the surrounding metal; this provides extra space for the rust expansion, but also creates structural weakness.

The process of corrosion and expansion generates an enormous force. For example, when Christopher Wren designed St Paul's cathedral, he specified the use of iron cramps; these were set in grooves so that their tops were flush with the surface of the stone blocks, but over the years the cramps rusted and expanded, exerting a vertical force on the layer of blocks immediately above. In the course of time the rusted cramps lifted and tilted the entire bell tower; this necessitated extensive repair work that involved removing the iron cramps and replacing them with cramps made from a stainless steel alloy. One metallurgist calculated that if all the available energy produced by the rusting ironwork in St Paul's could have been converted into useful work, it would have been capable of lifting a cathedral 1400 miles (2250 km) high!

What is acid rain?

It was a chemist called Angus Smith who coined the expression 'acid rain' in the 19th century. Smith was studying air pollution in Manchester, and found that rainfall had an acid content mainly in the form of sulphuric acid, dissolved in the water droplets after being emitted into the atmosphere as sulphur dioxide during the process of burning coal. Oxides of nitrogen and sulphur dioxide, as well as nitrate or sulphate particles, are oxidized in the air to form acids, and these can be returned to Earth through precipitation – rain, mist or snow – or as gases and dry particles which are absorbed into the surface.

Whereas sulphur dioxide is created by the burning of fossil fuels, nitrogen oxides are emitted by various industrial processes and vehicle exhaust fumes. European and American power stations that burn fossil fuels such as coal release about eight grams of sulphur dioxide and three grams of nitrogen oxides per kilowatt-hour. Acidity and alkilinity are measured in pH, on a scale of 1 to 14; a pH of 7.0 indicates neutrality, below 7 is acid, and above 7 is alkaline. In 1974, to illustrate the severity of the problem, acid rain fell in Scotland with a pH of 2.4; by way of comparison, the strong acid in a car battery has a pH of about 2.0.

Everything exposed to the air is potentially at risk from acid rain. Trees are particularly vulnerable. Sulphur dioxide enters the cells of a tree through fine crevices in the leaves, which are the centre of a plant's gas and water exchange system. In the cells, the sulphur dioxide reacts with water and turns into a substance poisonous to the plant. Nitrogen oxides, the main constituent of car fumes, have a similar direct effect on plant cells. In both cases, the damage shows up initially as a bleaching of the leaves. The poisonous substances acidify the soil and initiate a savage chain reaction. Nutrients are washed from the soil and beneficial minerals eroded, leading to the release of substances such as aluminium, which disturbs the absorption of important nutrients in the root system and makes it hard for the plant to 'breathe'. In addition, heavy metals are emitted into the atmosphere by industry and settle in the soil, where they accumulate after being transported by wind and rainfall. Some of these elements are so similar to the plant's chemical nutrients that the plant becomes confused and absorbs them instead, slowly poisoning itself.

In the 1980s, an international environmental survey examined 150 million hectares of forest throughout Europe. It was found that conifers were the worst affected. The problem was most serious in what was then West Germany, where a staggering 52 per cent of all forests were found to have suffered varying degrees of acid rain damage, caused mainly by industrial processes in Eastern Europe. Damaged trees exhibited yellowing and suffered loss of needles and foliage, and wood growth was also reduced. At the same time, there was evidence of damage in the root area, such as reduced growth of fine roots.

Acid rain also has an extremely harmful effect on buildings. Its sulphuric acid content transforms certain parts of stone – usually lime –

into sulphates, destroying the material's bonding action. Sulphate crystals crumble when soaked but form again as they dry out, so that a destructive potential remains in the stone. It would continue to be active, even if some day pollution were to be eliminated completely. Unlike living organisms, which have defensive powers to resist the attacks of pollutants and can even recover once the external agent is removed, the damage to building materials is permanent. Even the slightest trace of a pollutant will trigger and feed the destructive process.

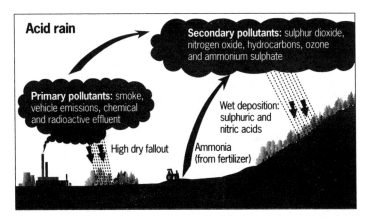

Acid rain

Secondary pollutants: sulphur dioxide, nitrogen oxide, hydrocarbons, ozone and ammonium sulphate

Primary pollutants: smoke, vehicle emissions, chemical and radioactive effluent

Wet deposition: sulphuric and nitric acids

High dry fallout

Ammonia (from fertilizer)

It has been proved that a quantity of sulphur dioxide, completely harmless to living organisms, will completely destroy natural stone in a few years if it is regularly present for a long enough period.

What is a mermaid?

Legends of mermaids – creatures that are half human female, half fish – are worldwide, and they are almost as old as recorded history. The mermaid legends of Europe and the Middle East probably originate in ancient Babylonia, where powerful fish deities were worshipped. These were associated with the Sun and Moon; the Sun was represented by Oannes, a god who had a human form but who wore a fish-head cap and a fish-skin cloak. In later Chaldean mythology he was replaced by Ea, a true fish-god who was half man and half fish; he eventually became the 'merman' of European folklore.

Greek mythology had the tritons, which were also half human and half fish; they had the ability to calm stormy seas and generally ruled the waves. Their airborne sisters were the sirens, who were half bird and half woman; their beautiful singing lured mariners to their doom. The Greek hero Odysseus got the better of them by plugging the ears of his crew with wax and binding himself to the mast so as to resist their blandishments. This potentially fatal singing has been an attribute of mermaids ever since, and was brought to a fine art by the Lorelei of the Rhine, as expressed in the traditional German folk song.

The very antiquity of mermaid and merman legends suggests that they may derive from a race memory of a time when the ancestors of the human race went through an aquatic phase of existence, a theory that is discussed elsewhere in this book. But this does not explain alleged sightings that have occurred frequently over the centuries, right into modern times. One theory is that such sightings were nothing more than wishful hallucinations in the minds of sex-starved sailors enduring voyages that sometimes lasted for months; it is perhaps significant that encounters with mermaids ceased almost completely when sail gave way to steam, drastically reducing the time spent at sea.

However, there may be more natural explanations. Mermaids have often been associated with seals, and in many legends the two exist in harmony. A tired sailor on watch might easily mistake a sleek seal form for a mermaid, especially if his head was already full of mermaid lore, and the cry of a young seal has an uncanny human quality.

Christopher Columbus, on his first voyage to the new world, reported having seen three mermaids leaping gracefully out of the sea off the coast of Guiana; what he undoubtedly saw were porpoise, and this gives another clue to the mermaid mystery. Sometimes, sailors on voyages into uncharted waters would sight sea creatures which are familiar to us today, but which were unknown then. Because they did not know what they had sighted, they would describe the creatures as mermaids.

One animal that has certainly given rise to mermaid stories is the manatee, a mammal of tropical coastal waters. About 8 ft (2 m) long, it has paddle-shaped front flippers and can rest upright on its tail, so that it seems to adopt a human-like stance. It has a powerful upper lip for grazing on aquatic vegetation, and a nursing female suckling her young has visible breasts.

Another sea creature with similar characteristics was Steller's sea cow, named after the German naturalist G.W. Steller, who discovered it in 1741. He identified it near Bering Island while searching for a route from Siberia to Alaska. It was much bigger than the manatee, with a length of about 30 ft (9 m). It had a small head and a large forked tail, and Steller observed that a pair of them mated like human beings, with a good deal of amorous foreplay. Large as it was, it could have been mistaken for a human-like creature when seen at a distance.

The confusion between sea cows and mermaids could not have lasted for very long, however. By 1768, the unfortunate animals had been killed off for their meat.

What is yeast?

It has been known for a long time that young and active plant cells, like those of the young grass leaf, are rich in protein. In the yeast plant, which is a microscopic fungus, the cells are constantly young, because they are perpetually renewing themselves by growing and dividing into new cells.

The word yeast, in fact, encompasses a variety of single-celled fungi, most of which belong to a class called *Ascomycetes*. They are asexual, and they reproduce by budding; a small bump protrudes from the parent cell, grows bigger, matures, and then detaches.

Yeast is 50 per cent protein and is a rich source of vitamin B, niacin, and folic acid. A deficiency in niacin causes pellagra, a chronic disease found mostly in subtropical countries where the staple diet is maize, while folic acid is essential for growth and the formation of red blood cells.

Yeast has been present in our food for a very long time, and when people were not too particular about clarifying their beer and wines they received the benefit of the valuable B vitamins in the yeast that turned their drink cloudy. Bread raised by yeast has its nutritional value improved by the proteins and vitamins of the yeast cell.

Fermentation and leavening are the principal uses of yeast in food manufacture. The fermentation process involves the breakdown of sugars by the yeast using the anaerobic method, which means that it does not require oxygen to secure a release of energy from a food molecule such as glucose. In other words, it does not need to breathe. The fungi can live on almost any sugary liquid, producing alcohol (ethanol) and carbon dioxide. The alcohol, of course, is the desired product in the fermentation process, while carbon dioxide is needed in baking. In the latter case, any alcohol produced is evaporated when the dough is baked. In beer and sparkling wines, some of the carbon dioxide is retained to produce a bubble effect.

What is the north-west passage?

When the Ottoman Turks overran the Middle East and Asia Minor in the middle of the 15th century, their conquests presented the crowned heads of Europe with more than a king-sized problem. At one stroke, the west's overland trade routes with China and the Indies were severed, or at best rendered so dangerous that no merchant would risk travelling over them.

It was for this reason that, in 1492, Christopher Columbus set out with his little expedition to seek a westabout sea route. Instead, he came up against the impenetrable barrier of the American continent.

Six years later, the Portuguese explorer Vasco da Gama succeeded in reaching India after a perilous voyage around Africa, and in 1521 his countryman Ferdinand Magellan made an even more hazardous voyage around Cape Horn to reach the East Indies.

Other explorers considered it possible to reach the Far East by a northerly route. The pioneer was the Dutch navigator Willem Barentz, who sailed from Amsterdam in 1594, rounded Norway to discover the island of Spitzbergen, and went as far as Novaya Zemlya before returning in 1597.

But it was the north-west passage – the opening of a sea route to the east via the Arctic seas to the north of the Americas – that attracted the imagination of British and French explorers. Their endeavours began in 1497, when King Henry VII despatched John Cabot to explore such a route to the Orient. He failed, and a number of subsequent expeditions met with disaster. In 1610, Henry Hudson thought he had found the way through, but his discovery turned out to be the huge land-locked and icebound bay that bears his name today. His crew showed their displeasure by casting him adrift, along with his son and seven others.

The knowledge that an Arctic passage to the east existed at all came slowly, through information built up over a period of two centuries or more by a succession of explorers and adventurers. In fact, the first transit of the North-West Passage was accomplished as the direct consequence of a tragedy.

In 1845, Sir John Franklin and 129 crew on the exploration ships *Erebus* and *Terror* vanished without trace in the Arctic. A search operation was mounted, and one of the searchers, Robert McClure, entered the icebound wastes to the north of Canada from the west. After two winters locked in the ice, he continued the journey by sledge to join up with a rescue vessel coming from the east. By that time it was nine years since Franklin had disappeared. The first true seaborne passage, however, was not achieved until 1906, when the Norwegian explorer Roald Amundsen – who was later to beat Captain Scott to the South Pole – completed the voyage in a 47-tonne converted herring boat, the *Gjoa*. It took him three years.

A commercial route from the Atlantic to the Pacific via the North-West Passage has yet to be opened. The route itself is well defined; it is located 500 miles (800 km) north of the Arctic Circle and consists of a series of deep channels through Canada's Arctic Islands. From a point north of Baffin Island it runs for about 900 miles (1450 km) westwards to the Beaufort Sea, north of Alaska. Entrance to the Pacific is gained through the Bering Strait, between Alaska and Russia.

The sea route via the North-West Passage is extremely hazardous because of ice and icebergs, and new icebreaking techniques would have to be developed before it could be exploited commercially. Nevertheless, it is a route that could be used by far larger vessels than those able to negotiate the Suez and Panama Canals, and its use would virtually halve the sea mileage between western Europe and, say, Japan.

What is soil?

Soil – or earth, to give it its other name – is a vital part of the triad of natural resources that are vital to our survival, the others being air and water. It is so important, in fact, that we named our planet after it. And yet, although we tread upon it every day, few of us have any idea of what it really is; even farmers, who may know lots about surface soils are often ignorant about what goes on deeper down, at the levels from which the crops they grow draw their nutrients.

Soil, basically, is like a blanket overlying the bedrock of the Earth's surface. It comprises a loose covering of broken rocky material and decayed organic matter; most of the latter is in the topmost layer, which on agricultural land is normally as deep as the bite of a ploughshare. Beneath this topsoil is the subsoil, consisting of mineral particles.

Soil is classified into several main groups, including immature soils, shallow soils, clay soils, well-drained brown soils, seasonally waterlogged soils, artificial soils, peats and podsols. The latter are light-coloured soils found mainly under coniferous forests and in cool regions where rainfall exceeds evaporation. In this type, the constant downward movement of water washes nutrients from the soil, making it unsuitable for agricultural purposes. Peat is a fibrous organic substance found in bogs, and is formed by the incomplete decomposition of plants in soil where there is a high level of acidity; this impoverishes the fauna and flora in the soil that would normally break down plant remains and incorporate them into the soil.

To obtain some idea of the composition of soil layers, you can try a simple experiment. Take some soil, put it into a glass jar with water, give it a good shake and allow it to settle. At the top of the jar, floating on the water, you will find a layer of humus, a dark-coloured soil component consisting of partly decomposed organic matter. Below this, suspended in the water, are fine clay particles; and at the foot of the jar, from top to bottom, is a layer of silt – a sediment whose particles have a coarseness midway between clay and sand – then a layer of coarse sand particles, and finally one of gravel.

Soil is subject to erosion, which is a natural process brought about by the actions of wind and water. Research has shown that in areas of grassland and woodland undisturbed by agriculture, there is an annual soil loss of about 0.1 tonnes per hectare. This loss, however, is cancelled out by the formation of new soils by weathering, the process in which exposed rocks are constantly being broken down by the effects of rain, frost, wind and other elements. In some cases, the volume of replacement soil can be as much as one tonne per hectare.

This, unfortunately, is not the case on agricultural land, where topsoil erosion is severe. According to some estimates, there will be a third less topsoil throughout the world by the beginning of the 21st century than there is now. Half of America's farmland is losing topsoil faster than it can be replaced; the loss amounts to 1700 million tonnes every year.

The problem is that the ever-growing demand for food production has led to an increase in the size of fields, which then become catchments for water trapped on the surface of the soil before being compacted by agricultural processes. The problem is particularly acute in tropical climates, which are subjected to heavy seasonal rainfall.

In reality, it all amounts to a vicious circle. Soil is essential to mankind's continued survival, but the more we exploit it, the less of it remains to produce the crops we need.

What is an avalanche?

This sounds like a very simple question. After all, everyone knows about avalanches – the tumbling walls of snow that can cause devastation and loss of life in mountainous areas. In fact, it is only recently that scientists have really begun to understand what causes an avalanche.

Although it is difficult to be precise, because many of the worst avalanches occur in remote parts of the world where casualty figures are not recorded, the snow deluges are thought to kill several hundred people every year. The greatest avalanches occur in the Himalayas and Andes, and in the latter case sometimes cause terrible loss of life. In December 1941, for example, an avalanche wiped out the town of Huaras in Peru and killed an estimated 5000 people.

In Europe, the worst spate of Alpine avalanches occurred in Switzerland, in 1720, a particularly severe winter when dozens of villages were swept away and hundreds of lives lost. In the First World War, at least 40,000 Austrian and Italian soldiers on the Alpine Front are thought to have perished when the snow came plummeting down on them. In 1885, an estimated 120 million cubic ft (4 million m³) fell in one avalanche in the Italian Alps.

It is now known that there are three different kinds of avalanche. A soft-slab avalanche takes place when a new layer of snow breaks away and starts to slide. This quickly turns into a huge cloud of powdered snow that rushes downhill at 100 mph (160 km/h), pushing a wall of wind ahead of it. In fact, it is the wind that causes most of the destruction. About 80 per cent of all avalanches are of the soft-slab kind and usually occur within 24 hours of a blizzard, before the surface snow has had time to become compacted.

The second type is the hard-slab avalanche, which consists of old, dense snow hammered into a hardened mass by freezing high-speed winds. When the top layer breaks away, it fragments into chunks that can be as big as a car. As hard as concrete, they travel down the slope at up to 50 mph (80 km/h), smashing through everything in their path.

The third kind, and the least dangerous, is the wet-slab avalanche, which usually occurs when liquid water – the aftermath of rain or a spring thaw – mixes with the snow. A wet-slab avalanche flows down a mountain slope at no more than 10 mph (16 km/h), giving plenty of warning of its approach.

The popularity of winter sports has led to a greater demand for efforts on the part of physicists and meteorologists to study – and perhaps some day be able to control – avalanches with the help of advanced computer models that monitor and predict weather patterns over the areas likely to be affected. Modern meteorological data warns resort managers when avalanches are likely to occur, so before the start of a day's skiing they create their own controlled avalanches. This is done by placing small explosive charges at strategic points. Charges of TNT may also be dropped from helicopters. This, however, is a costly business, so major resorts tend to use a new weapon in the battle against the avalanche – a cannon powered by compressed nitrogen that can hurl a two-pound charge over 2000 yards.

One result of avalanche studies has been the creation of special zones. No new buildings may be erected in Red Zones, where major avalanches have been known to occur within a 25-year time scale; and in Blue Zones, where weaker and less frequent avalanches are predicted, structures must be able to resist any weight of snow that might slide down on them.

Modern technology is also coming to the aid of skiers who have the misfortune to be caught in an avalanche. Many skiers now carry miniature transmitters enabling them to be located under snow, and in the latest development a skier wears a diode with a foil antenna between the inner and outer layer of his boot. This produces a radar reflection that can be detected by equipment carried in a rescue helicopter.

What is a Komodo dragon?

In 1912, a Dutch aviator – one of the pioneers of early aviation in what was then the Dutch East Indies – made an emergency landing on Komodo, an island in the Lesser Sunda group off the coast of Java. The man was rescued, and returned to civilization with horrifying stories of having encountered giant, lizard-like creatures. Most people dismissed the tale as preposterous, but the curator of the Botanical Gardens on Java was sufficiently interested to ask the Dutch Civil Administrator of Nusa Tenggara Timor Province, to which the Lesser Sunda Islands belonged, to investigate further. A preliminary expedition came back with the skin of a 7 ft (2 m) reptile and assurances from the natives of Komodo that a fully-grown specimen often reached 30 ft (9 m) in length.

A second expedition was mounted, led this time by a Malay hunter who specialized in tracking down elusive animals, and a local rajah provided other trackers and hounds. The hunters were in luck; they succeeded in capturing four huge lizards, the biggest of which was 10 ft (3 m) long.

Classified as *Varanus komodoensis*, the lizard was identified as a species of the monitor lizard family, and became known popularly as the Komodo dragon. Subsequent research showed that the creature's maximum length is about 10 ft, so the Komodo islanders had exaggerated a little; they may have been misled by the fact that it digs a 30-ft burrow in which to lay its eggs. It weighs about 300 lb (135 kg) and has a life span of up to 100 years.

The newly hatched young are about 18 in (45 cm) long and live in trees for several months while they mature – a necessary precaution, because adult Komodo dragons are cannibals and will eat smaller members of the species and occasionally other adults. The tree-dwellers drop on their prey, and adults are swift runners; they have sometimes been known to attack and devour human beings. In general, though, their main diet is carrion, and adults may travel several miles a day from their burrows in search of food.

Over the years the Komodo dragon has been hunted almost to extinction by collectors, and today it is a protected species. It may have a relative in Papua New Guinea, where tribesmen speak of a creature called *Aou-Angi-Angi*, 'the crocodile that lives in trees'. It is mentioned in *The Annual Report of the Territory of Papua for the year 1936-37*, by the then Lieutenant-Governor, Sir Hubert Murray, who wrote:

'I first heard of a "land crocodile" or "tree alligator" many years ago. I have not seen one, and its existence was doubted. Ahuia-ova, a well-known native of Port Moresby, also told me that he had shot one of them, and seen two others. The one he shot had been fighting a pig which it tore in pieces. It was, he says, as big as a small crocodile...The other two which he saw were, he said, so big that he was afraid to shoot; in one case he hid behind a tree, in the other he ran away. Ahuia had a gruesome story to tell of a Gorohi native whose dog was seized by one of these "alligators". The Gohori saved the dog, but was seized himself, carried up a tree, and torn to pieces.'

There is little doubt that the forests of the Indonesian archipelago are impenetrable enough to hide such creatures, and others may also exist in various remote parts of the world. Australia's trackless wastes may provide ideal cover for creatures of which scientists as yet have no knowledge of, other than that hinted at by aboriginal stories; but prospectors in the Outback have returned with claims of having sighted giant monitor lizards 30 ft (9 m) or more in length.

What is smog?

In December 1952, a lethal pall of smoke and fog – popularly called 'smog' – descended on London. England's capital was no stranger to the phenomenon; after all, reeking, dense fogs were part of the Victorian world of Charles Dickens and Jack the Ripper, and had been ever since the 17th century, when coal became more readily available as a domestic fuel. The smog of 1952, however, was different. It lasted for most of a week, and in that time 4000 people died, four times more than the December average for London. Nothing like this had happened before.

Smog is a mixture of smoke and sulphur dioxide suspended in moist, foggy air. London was not the only city to suffer from it, but because it was a congested city in the 19th century its effect was more noticeable. In the first half of the 20th century air pollution became less concentrated as the capital expanded, but after the Second World War dense winter smog once again became a nuisance, caused by the burning of low-grade coal – and, even worse, by briquettes, which were a noxious mix of coal and cement dust – a necessity created by severe post-war fuel shortages (the cement slowing down the burning process).

On this particular occasion in 1952, London was subjected to a temperature inversion. Normally, air close to the ground is warmer than the air above it, and tends to rise; but a temperature inversion results when the surface temperature falls rapidly enough to chill the layer of air above it, causing mist to form as dust particles attract water vapour. Normally, this inversion is rapidly dispersed by the rising sun, whose rays penetrate the mist or fog layer and heat the surface, which in turn radiates heat to the layer of air above it. In this case, however, the accumulation of smoke near the ground was so dense that sunlight failed to penetrate it, so the lower layer of air remained freezing and static. The smog was yellowish-black and stank of sulphur; pollution built up on pollution, and at its worst the 'pea-souper' reduced visibility to between one and five yards, extending for 19 miles (30 km) around London. At the end of five days, to the relief of everyone, a westerly wind blew it out into the North Sea. But in its wake it left 4000 fatalities, mainly the very young, the very old and people suffering from various forms of respiratory ailments.

The smog resulted in the government of 1956 passing the Clean Air Act, authorizing local councils to set up smokeless zones and to provide grants to householders for the conversion of their homes from coal fires to other forms of heating, such as gas and electricity. It was a slow process, and it did not entirely solve the problem; London smogs still occur, but mostly in summer. Nowadays, they are the products of temperature inversions coupled with the nitrogen oxides emitted by vehicle exhaust fumes. Sunlight actually exacerbates the effect, producing chemical reactions that create secondary pollutants. The overall result is a chemical cocktail that is extremely harmful to sufferers from respiratory complaints, particularly asthma.

In some of the world's most overcrowded cities, the effects are far worse. Los Angeles is a prime example, as is Tokyo; both suffer severely from photochemical smogs as sunlight acts on the range of pollutants ejected into the air. Some, like Mexico City, are pollution disasters; Mexico City is the most populous capital in the world, with 20 million inhabitants and three million vehicles. Its pollution problems are increased by its altitude of more than 6000 ft (2000 m) above sea level; oxygen pressure is 20 times less than at sea level, so fuels burn less efficiently and discharge more pollutants.

What is malaria?

Worldwide, malaria kills more people than any other disease. It is one of the oldest diseases known to mankind, and probably wrought havoc among struggling prehistoric populations. The parasites that cause it are carried in the bodies of 60 different species of mosquito and are transmitted to humans by the female insect.

Male mosquitoes are not equipped to bite; their principal food is plant nectar. So is the female's, but she needs a 'fix' of blood once or twice in her lifetime to provide essential vitamins. If she does not get it, her breed begins to weaken and eventually dies out. With the blood she takes up malarial parasites, assuming the victim has the disease.

The problem for humans starts with an infected female mosquito. Before she draws blood from the capillaries, she unblocks her proboscis by injecting hundreds of microscopic parasites into the victim's bloodstream. Within half an hour these find their way into the liver, where they are safe from counter-attack by the body's defensive system. In the liver the parasites multiply and then re-enter the bloodstream in a modified form, invading the red blood cells. Infected cells eventually rupture and release more parasites which infect other cells throughout the body in a kind of chain reaction.

Symptoms – alternate periods of shivering and sweating, sickness, body pains and so on – usually start to develop within a week to 14 days. However, one type of parasite – *vivax malaria* – can lie dormant in a person for years. Malaria attacks today are not uncommon among men who fought in the Far East during the war, and still more frequent among those involved in the post-war campaigns in Malaya and Borneo.

The first effective treatment for malaria was identified in 17th-century Peru, where Indians used the bark of an evergreen 'fever tree' to concoct a remedy. The new bark cure was taken to Rome in 1632, probably by Jesuits, and it was known in England by 1658. Oliver Cromwell caught the disease, but his doctors considered the 'Popish powder' an unfitting medicine for him, so he went on suffering.

A few years later a London physician named Robert Talbor achieved fame and fortune by using the bark of the cinchona 'fever tree' to cure such notable malaria sufferers as the Queen of Spain, but it was not until 1820 that two French chemists isolated an alkaloid which they named quinine from the cinchona bark. Cinchona forests were later planted on Java, which eventually produced most of the world's supply of quinine, but when the Japanese overran the Dutch East Indies during the Second World War this vital source was cut off, compelling British and American scientists to search for synthetic substitutes. They came up with an effective anti-malaria drug called chloroquine, and the pesticide DDT, which attacked the mosquitoes themselves. Used together, these brought about the virtual elimination of the disease from some badly affected countries in the post-war years, and led to real hopes in the 1960s that malaria could be eradicated altogether.

Then the problems started. First, the mosquitoes became resistant to DDT, and then they began to develop resistance to chloroquine also. The result, in recent years, has been a great upsurge in the disease throughout the tropics, with a possible threat to parts of the USA and southern Europe. In India, for example – one of the worst affected countries – malaria cases average around 60 million a year.

Africa, already torn by drought, famine, and a host of other troubles, is the worst hit. According to one World Health Organization estimate, malaria kills one million African children under the age of five every year – and the total number of cases on the African continent may exceed 250 million annually.

So the race is on again to find a new drug in the war against malaria. One of the latest to be used in the field is melfloquine, developed by the US Army during the Vietnam War – but already, some species of mosquito are reportedly developing resistance to it. Another problem with malaria treatment is that the symptoms – at least in the early stages – resemble those of influenza. This can lead to a loss of several vital days before a sufferer receives proper treatment.

The real answer lies in developing an anti-malaria vaccine, but this could be a long way off. In the meantime, mankind is barely holding its own in its oldest battle with the insect world.

What is sleeping sickness?

Sleeping sickness is the scourge of tropical Africa. Like malaria, it is caused by a parasite delivered through the bite of an insect. The single-celled (protozoan) parasite in this case is *Trypanosoma brucei*, and the disease it causes is also known as African trypanosomiasis.

The creature responsible for transmitting the organism is the tsetse fly. The name *tsetse*, in the languages of Africa where the creature proliferates, can mean simply 'fly', or 'fly that destroys cattle'. There are 21 species of it, and it is a relative of the housefly, although a little larger, and varying in colour from yellowish brown to dark brown. It has a life span of up to three months, one to two months being more usual.

Like the mosquito, the tsetse fly needs animal blood for survival, but it is a voracious feeder, sucking blood almost daily. Another parallel with the mosquito is that without an adequate supply of blood, the female fly will produce a larva – which hatches from an egg within her – that is small and underdeveloped. Under favourable feeding conditions, she produces a fully matured larva almost every day of her adult life. The larva immediately burrows into the ground and turns into a pupa, emerging as an adult fly after several weeks.

Around 80 per cent of the tsetse flies that attack humans are males; the females generally attack larger animals, their parasites producing a disease called *nagana*. This is similar to the sleeping sickness that affects humans, and occurs mostly in East Africa. In West and Central Africa humans are the main victims, and the disease is spread from person to person.

In both forms of the disease, a painful nodule develops at the site of the tsetse fly bite, but subsequent development is different. In the East African form, a severe fever develops within a few weeks of infection, and the victim may die of heart failure before the disease reaches the brain. In the West African form the disease runs a slower course; the victim experiences recurring bouts of fever accompanied by enlargement of the lymph nodes, the small lumps of tissue that occur in the lymphatic system, and after months or even years the disease spreads to the brain, causing headaches, confusion and eventually severe lassitude. The victim, if the disease goes to its extreme, becomes completely inactive, with a vacant expression and drooping eyelids. Without treatment, coma and death are the inevitable outcome.

The disease can be countered, and in most cases a complete cure can be effected, although there may be residual damage if the disease has spread to the brain. As is the case with malaria, the earlier the disease is diagnosed and treated, the greater the chances of recovery. Drugs such as tryparsamide, an arsenic-based compound, are highly effective but they may produce severe side effects, even loss of vision through acute optic neuritis in extreme cases. In the Cameroons, infection was reduced from a high percentage to less than one per cent by treating all diagnosed cases with tryparsamide, a measure that saved entire tribes from extinction.

The most effective controls, however, have involved destroying the fly's environment by clearing woodlands, preventing the growth of brush through regular burning, and destroying the wild life that is the insect's primary source of nourishment. This wholesale destruction of the environment may be unpalatable to many, but it appears to be a necessary evil. Despite all the other forms of offensive that have been launched against it over the years, the tsetse fly stubbornly refuses to be wiped out.

What makes a rattlesnake rattle?

The name 'rattlesnake' covers a multitude of reptiles – any pit viper, in fact, that belongs to the classification *Crotalus* and *Sistrurus*. There are 47 species in North and Central America, concentrated mostly in northern Mexico and the southwestern USA. The sidewinder, less than 30 in (76 cm) in length, is one of the smallest species, while the eastern diamondback, which lives in the area from North Carolina to Louisiana, is the heaviest living poisonous snake, growing to 7 ft 3 in (2.2 m) and weighing 15 lb (6.8 kg). On average, the western diamondback, which occurs from southeastern Missouri to Baja California, is a little smaller, but it has been known to reach 7 ft 5 in (2.3 m) in length, while the banded or timber rattlesnake grows to 6 ft 2 in (1.8 m).

What makes rattlesnakes unique among other reptiles is their rattle, which is composed of loosely interlocking segments of horny material; these produce a sharp rattling sound when the tail is vibrated. The snake acquires its first permanent rattle when it is about ten days to two weeks old, at which point it sheds its first skin; after that, moulting takes place two to four times annually, and with each moult a new rattle segment appears at the base. Sometimes, the age of a rattlesnake can be roughly estimated by dividing the number of segments by three, but this is not an altogether accurate method because segments break off or are gnawed off by rats when the snake is hibernating.

The notion that a rattlesnake rattles its tail just before striking is a myth. A snake will almost invariably rattle as a warning signal if approached out in the open, but if it has some form of cover – a log or bush, for example – it may lie perfectly quiet, depending on silence for protection. Rattlesnake bites do occur, but they are usually as the result of an accident, such as when a basking snake is trodden on. In general, although their habits vary from species to species, rattlers are much more timid than they are aggressive. If the sinister-sounding rattle does not halt the approach of an enemy, the snake raises the forepart of its body into a threatening S-shaped curve, but it will only strike as a last resort.

Rattlesnakes feed primarily on rodents and can live for about a year without eating; this is about the maximum length of time a rattlesnake will survive in captivity, for after capture it usually refuses food. A rattlesnake is less dangerous after it has eaten because it expends about two-thirds of its venom in killing its prey, and replenishing the supply takes a couple of weeks.

Female rattlesnakes bear anything between three and 24 living young, some – depending on the species – having a brood only every other year. The young possess well-developed fangs and begin hunting – usually for field mice – within a few weeks of birth.

The rattlesnake is one of nature's most friendless creatures. Humans kill them to protect livestock – and often just for fun – and collect them for the production of antivenins. Many animals and birds kill and eat them; even chickens gobble up baby rattlers, just as they would worms. All in all, it's not a rattler's world. In fact, it is something of a miracle that they continue to survive.

What is crying?

The human eye has a very efficient cleansing system called the lacrimal apparatus, which produces the salty, watery solution we call tears. Its primary function is to keep the cornea and conjunctiva (the transparent membrane covering the white of the eye and the inside of the eyelids) constantly moist, because moisture is essential to maintain the transparency of the cornea and prevent ulceration.

The eye's main lacrimal glands secrete tears during crying and also when the eye is irritated; these glands lie just inside the upper and outer margin of the orbit – the socket in the skull that contains the eyeball, protective pads of fat, and various muscles, nerves and blood vessels – and drain into the conjunctiva. The latter contains accessory lacrimal glands which secrete fluid directly on to its surface.

The tear film over the cornea and conjunctiva consists of three layers: an inner, mucous layer secreted by the glands in the conjunctiva, a middle layer of salt water, and an outer, oily layer secreted in glands (called the meibomian glands) on the eyelid. In carrying out their necessary function of lubrication, tears assist the movement of the eyelid in blinking and wash away small foreign bodies; they also contain a natural antiseptic called lysozeme.

Tears sweep across the eye and drain away through tiny openings called the lacrimal puncta, which are situated towards the inner part of each eyelid. These are connected by narrow tubes to the lacrimal sacs, which lie in shallow hollows in the lacrimal bones, situated on either side of the nose just within the inner margin of the orbit. Flat muscles that overlie the lacrimal sacs compress them during the action of blinking, and this process of compression and release drains away excess fluid via the nasolacrimal ducts, which run down the bone and open inside the nose – the reason why a runny nose accompanies a bout of crying.

One of the main functions of crying, in human beings, is to express emotion. This is a throwback to our infancy, when crying is the only means at our disposal to express an urgent need or

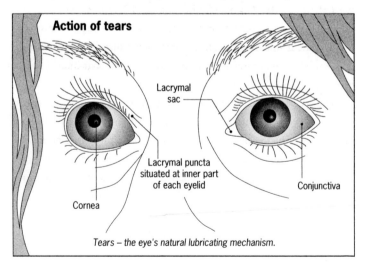

Action of tears

Lacrymal sac

Lacrymal puncta situated at inner part of each eyelid

Conjunctiva

Cornea

Tears – the eye's natural lubricating mechanism.

discomfort – hunger, thirst, a soiled nappy or an overwhelming desire to be comforted. Crying is also an excellent means of relieving stress, and persistent crying is an indication of depression.

So why do people cry when they slice onions? Quite simply, onions and garlic both contain derivatives of sulphur-containing amino acids, some of which are decomposed by an enzyme during the cutting process to form propanthial sulphoxide, a powerful irritant which produces propanol, sulphuric acid and hydrogen sulphide when it comes into contact with water – in this case, the film of tears covering your eyes. The eyes try to dilute the acid by producing more tears. There are all sorts of old wives' remedies for this, but the simplest method is to keep the onion moist as you slice it.

What is a laser?

The laser – the word is an acronym for Light Amplification by Stimulated Emission of Radiation – was a logical progression from the maser (Microwave Amplification by Stimulated Emission of Radiation), the principle of which was first proposed independently by American and Russian scientists in 1954. A maser is a high-frequency microwave amplifier or oscillator, in which the internal energy of atoms is used to obtain low noise-level amplification and microwave oscillations of precisely determined frequencies. Stimulated emission, the underlying principle behind both masers and lasers (which are sometimes called optical masers) occurs when an unstable atom is stimulated into emitting energy at the same frequency.

Microwaves, radio waves and X-rays are some of the many

manifestations of electromagnetic radiation; light is another. When light is emitted or absorbed by an atom, single quanta (discrete units) are also emitted or absorbed; each atom emits or absorbs only quanta of specific wavelength, which is determined by the energy of the quanta. The wavelength in turn determines the colour of the light; a good example is the fiery red glow produced by neon gas used in electric advertising signs.

Neon is just one of many gaseous, solid and liquid substances that have been used in lasers; in fact, any substance whose atoms can be put into an excited energy state can be used for laser material. This material is called the 'active medium', in which most of the atoms can be 'pumped' to an excited state by subjecting the system to electromagnetic radiation of frequencies which differ from that of the stimulating frequency. In the case of a laser, the atoms are stimulated by the presence of light in a process known as stimulated absorption and stimulated emission. Stimulated absorption decreases the number of light quanta, so that the intensity of the light diminishes but the atoms gain energy; stimulated emission, on the other hand, produces two light quanta where only one existed before, and the atoms gain energy as a result. If the net stimulated emission in an assembly of atoms exceeds the net absorption, laser action takes place.

The active medium used in the first laser, developed in 1960, was a synthetic ruby crystal, the ruby being in the form of a cylindrical rod with both ends silvered to act as mirrors. One end is more lightly silvered so that some light can pass through it. In this type of device, laser action begins when light from a flash tube initiates an exchange of energy within the chromium that forms a small part of the ruby's chemical composition; the chromium atoms are raised from their ground (stable) state to broader energy levels where a considerable range of wavelengths of the stimulating light is effective in producing excited atoms. Once the chromium atoms have reached these levels, they quickly pass into a lower energy state; however, the energy lost in this process is not radiated, but is released directly into the aluminium oxide that forms the rest of the ruby's structure, where it is eventually dissipated as heat. The quanta emitted in this process are reflected to and fro between the silvered ends of the ruby rod, and this stimulates other excited atoms to release their quanta, gradually building up the energy that forms the laser pulse.

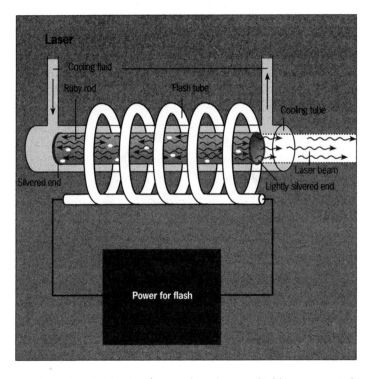

The pulse of light that forms a laser beam is highly concentrated, and coherent. Its rays are nearly parallel to one another, and diverge only slightly as they travel. During an experiment carried out in 1962, a laser beam 1 ft (30 cm) in diameter illuminated an area of the

Moon's surface 2 miles (3 km) across; by way of comparison, a beam of ordinary light travelling the same 240,000-mile (386,000 km) distance would illuminate an area 25,000 miles (40,000 km) in diameter.

Lasers today are among technology's most useful tools. They have revolutionized certain areas of medical science, particularly in treating retinal damage; they are widely used in communications, photography, ranging and surveying, and weapons guidance, to name only a few applications.

Index